LIFE ON
OUR PLANET

LIFE ON OUR PLANET

DR TOM FLETCHER

sourcebooks

CONTENTS

THE GREATEST STORY EVER TOLD
IS WRITTEN IN STONE.

THE EARTH BENEATH OUR FEET
RECORDS BILLIONS OF YEARS OF HISTORY
AND IS A MONUMENT TO THAT
UNIQUELY COMPLEX AND FRAGILE
CHEMISTRY WE CALL LIFE.

USING THE LATEST TECHNOLOGY
AND SCIENCE TO BRING LONG
EXTINCT CREATURES BACK TO LIFE,
LIFE ON OUR PLANET TELLS
THIS INCREDIBLE STORY.

Our galaxy, the Milky Way, is just one among a trillion others in the Universe. It alone may contain 400 billion stars, scattered across its 830-quadrillion-kilometre span. Drifting humbly within it is our solar system and our planet, Earth, orbiting at just 93 million miles from the sun. In our tiny plot of the Universe, a stroll around Earth is nothing, at around 40,000 kilometres. To even intelligent apes like us, these measurements are far beyond the scope of what we need to comprehend in order to survive. Numbers over a thousand – above a tangible quantity – quickly become abstract and meaningless. Equally incomprehensible is the antiquity of the Universe – 13.8 billion years old. The solar system and Earth have existed for just a third of this time, 4.54 billion years. What followed was a spectacularly fortuitous chain of events, that as far as we know is utterly unique. There was a transition from stark and simple chemistry to biology and the evolution of life itself surprisingly soon thereafter – a mere 4 billion years ago. It was the start of an incredible story that we are only beginning to understand.

Our own species, *Homo sapiens,* is only 300,000 years old, so we have been around for less than 0.007 per cent of Earth's existence. These great tracts of time are uncomfortable to imagine, especially as they highlight the relative insignificance of humans

RIGHT: A herd of woolly mammoths soldier forward through a long-lost ice world. Once mighty animals, mammoths are now potent symbols of extinction.

in the planet's long history. However, of all the species that have ever lived, we are the first to have learned about our place in time and space. Thanks to our curiosity and intelligence, we are in a unique position to tell our own story, in the knowledge that we are not here by accident. We are the last of a long line of survivors who, by virtue of our own existence, we know all lived to reproduce. Each individual organism between us and the very first was a winner because its genetic blueprint made it to the next generation. Indeed, every single life form alive today is just the latest of countless iterations of a biological formula that worked.

Earth today is home to at least 10 million species, and some scientists estimate there may even be billions – perhaps over a trillion. Even in one gram of soil, you could find 50,000 species of bacteria if, of course, you cared to look. However, this is nothing compared to what has been before, with over 99 per cent of the species that have existed now extinct. The legacy of these fallen hordes is the genetic code baked into their descendants' DNA and scraps of their remains, which lie fossilised in the rocks. Fossils are shadows of the life we see around us today and the ultimate cold case for scientists trying to uncover their secrets. Extinct creatures have always captured our imagination, from the weird and wonderful life of the ancient seas to the giant dinosaurs that paced the land. Our museums are filled with these remains, but the rocks that hold them are also rich with information, recording incomprehensibly vast amounts of time.

We are living in an exciting era of scientific discovery, having only learned to read the language of these rocks in the last century or so. Kilometres of these layers lie beneath our feet – a record of evolution, geography, climate and crisis. Over hundreds of millions of years the relentless shifting movement of the Earth's crust has distorted and warped these rocks. Tectonic plates constantly reshape the planet's surface, thrusting mountains skywards, driving continents across the globe and causing entire seas to rise or drain away. From our own perspective these processes are so slow that it is difficult to see that they are happening at all. That is, until we are reminded of the planet's violence by earthquakes or volcanic eruptions. We can see that the landscape is covered in the scars of these great tectonic contortions, and one fortunate result of these events is that even the deepest and oldest rocks can be pushed up to the surface for us to find.

In the past few decades alone, scientists have made immense progress in piecing together the Earth's story, using all the analytical tools available to them. Scarcely a week goes by without a new fossil discovery, and much has changed since the old textbooks were written. Innovations in technology have also revolutionised the way fossils are analysed. Traditionally a fossil would be prepared for months, if not years, behind closed doors by carefully rasping away the rock

RIGHT: Humans are a uniquely intelligent force. Our history, and the history of our planet, can teach us much about surviving in balance with the Earth.

that entombs it. Today, within minutes a specimen can be scanned in beautiful detail, providing a three-dimensional virtual model, which, within a few minutes more, can be shared with colleagues all over the world. Increased computing power means that palaeontologists have all manner of tools at their disposal. Thankfully, these are becoming far more accessible in newly industrialised countries, where research groups are growing in order to study their own natural heritage. Palaeontology today is a truly interdisciplinary science, involving biologists, ecologists, chemists, zoologists and even engineers. More than ever, palaeontologists are embracing collaboration and shifting away from the more traditional way of working, which involved individuals harvesting treasures from another continent and describing them in secrecy.

This revolution in science means that we know more than ever about these lost ancient worlds that capture our imaginations. *Life On Our Planet* follows life's grand journey from the chemical and biological soup of its simplest times to the majesty and incredible diversity of life on Earth today. Using cutting-edge camera techniques, expertly crafted visual effects and a wealth of new scientific information, the series is a whistle-stop tour of the greatest dynasties ever to live. This book will dive even deeper into this abundance of new knowledge and understanding, building on the themes and stories featured on screen. Both should be seen not only as a biography of life, but also an opportunity to reflect on the lessons we can learn from the deep past. Just as we shine the light of discovery on our biosphere, we are also destroying it with ruthless efficiency. Throughout the hundreds of millions of years in which complex life has existed, it has faced a number of lethal challenges, which have verged on total annihilation. By learning from prehistory, we may not be doomed to repeat it.

LEFT: *Lystrosaurus,* a small but miraculous survivor of Earth's greatest mass extinction, 252 million years ago.

ORIGINS

Before the curtain rose for the greatest show in history, history itself had to begin. About 13.8 billion years ago a singular point, no bigger than the dot at the end of this sentence, became a colossal Universe in spectacular fashion. It expanded extremely rapidly, maybe faster than the speed of light, to create space, matter and antimatter, fundamental forces like gravity - in fact, everything. After 300,000 years of unimaginably energetic expansion, atoms could begin to form and a long road to calm began.

Our solar system formed only 4.57 billion years ago, when swirling clouds of cosmic dust and gases gradually coalesced into solid structures. Over hundreds of millions of years, lumps of molten rock collided and took shape, drawn together by gravity. In the early chaos, some were slung out to the cosmos, while a handful rested in orbits around our star, the Sun. As planets grew bigger, they pulled ever-larger pieces of debris towards them and, eventually, collapsed into spheres.

This was a time of instability and intense astronomical violence, where chance could decide the fate of an emerging world. As the protoplanets became larger, so too did the impacts, and towards the end of its formation, Earth itself came dangerously close to annihilation. Around 4.5 billion years ago, an enormous Mars-sized body called

PREVIOUS PAGE: Dazzling pools of extreme chemistry - seen here in Dallol, Ethiopia - test the limits of where life can exist.

RIGHT: Earth is unique in the Universe, cradling all life that has ever lived and orbiting one star nestled among 400 billion others in the Milky Way.

Theia collided with Earth, pulverising everything in its path, and remelting its solid interior, the mantle. Amidst this explosive chaos, debris broke away from Earth to form our Moon, which has remained tethered in our orbit ever since.

Throughout the formation of the solar system, the Sun was at the centre of this cosmic dance and its gravity and heat would have a huge influence on the planets around it. Denser materials, such as metals, were naturally drawn closer to the Sun by gravity, whereas lighter gases collected further afield. As a result, the inner four planets (Mercury, Venus, Earth and Mars) emerged with iron-rich cores and rocky solid surfaces, and planets beyond were left with more volatile and lighter chemistry. As the solar system cooled, some planets developed atmospheres, enriched with volcanic gases such as methane, carbon dioxide and water vapour. These atmospheres helped to trap solar radiation, creating a greenhouse effect that elevated the planets' surface temperature beyond what it should be with the Sun's direct heat alone. Venus, for example, should be frozen but it is, in fact, the hottest planet in our solar system. The cause is its own atmosphere, which contains 154,000 times more carbon dioxide than Earth's. The greenhouse effect there is so severe that the oceans of Venus have boiled away and lead would melt on its surface.

RIGHT: A lava flow near Kalapana, Big Island, Hawaii. Outpourings of lava and greenhouse gases sculpted the Earth's first landscapes and defined its chemistry.

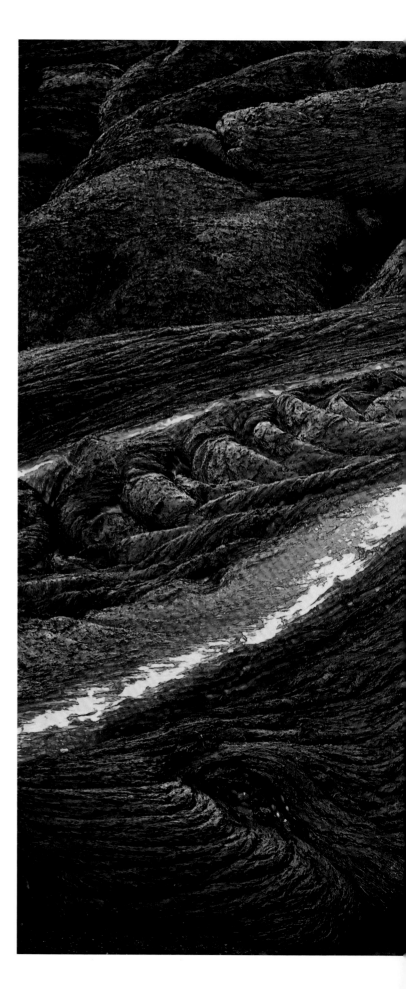

Earth, too, should be completely frozen, but fortunately the levels of greenhouse gases in our atmosphere have elevated the average temperature by 33°C. Crucially, this has allowed liquid water to exist on our planet's surface, and the Earth's huge gravitational pull (compared to, say, that of the Moon) has largely kept it here. A perfect storm of chemistry, astronomy and a little chance had made early Earth a warm and fertile water world, but the show had only just begun.

EARTH – BUT NOT AS WE KNOW IT

Though the stage was set for life to appear, the early Earth of 4.4 billion years ago would have resembled a hellscape. There was intense volcanic heat bursting from beneath the planet's surface, radiation from unstable elements peppering the unbreathable air, and a barrage of meteorites raining from the heavens. The skies would have been unrecognisable, hazed by methane and clouds of volcanic gases. No mountains had formed yet, so the only features on the Earth's surface were black volcanic arcs and the rims of impact craters. Indeed, the name given to this first time period in our planet's history is the Hadean, after the Greek god of the Underworld.

Shallow seas covered most of the Earth's surface, but only as a thin film of warm water, stained an unappealing iron-sulphide green. Remarkably, this water would be the key to every aspect of the history of life to come, and its first act would be saving the climate from a volcanic greenhouse. As it was exposed to the air, the water absorbed enormous amounts of carbon dioxide. This reacted chemically to become heavy carbonate minerals, which sank to the sea floor. The early seas ensured that excess carbon dioxide was buried and locked away as rock, preventing the runaway greenhouse effect that we see on Venus.

Water is crucial for facilitating chemical reactions, and without it life itself would not exist, but the origin of Earth's water is still hotly debated. Some suggest it has always been here and is a part of the planet's original chemistry, while others suggest early Earth was too hot to retain it. While still a controversial idea, it may be that the water was delivered here by meteorites, particularly a type called carbonaceous chondrites. As well as minerals and organic matter, these meteorites can contain over 20 per cent water, and they likely held much more than this four billion years ago when they collided with Earth. Around 3.9 billion years ago, a particularly intense bout of impacts affected the inner solar system, during a period called the Late Heavy Bombardment. Some of these asteroids – perhaps 40 kilometres wide – hit the Earth, Moon and other inner planets with enormous force, reshaping the primordial landscapes.

LEFT: Waves crash on a black sand beach. At first our planet resembled an alien world, but it had a crucial ingredient for life – liquid water.

With this doom and destruction, water — a vital ingredient for life — was being delivered, and some scientists have even suggested life itself could have hitched a ride. While an extraterrestrial origin for life is controversial, we do know that all comets and some meteorites contain organic molecules, including the building blocks of proteins. As exotic compounds rained down from the skies, ultraviolet radiation drove the formation of ammonia, carbon monoxide, formaldehyde, methane and more. In this planet-sized chemistry set, even cyanide was produced in vast quantities, a crucial building block for complex molecules such as amino acids and nucleic-acid bases. It is amazing to think that from this poisoned chalice of a planet, life would emerge.

WHAT IS LIFE?

A definition for life is tricky to pin down, but to qualify most biologists agree it has to have a flow of energy in organised structures, and the ability to replicate itself. In effect, life is a system of chemical reactions, bounded and protected from the outside world by membranes. Before true life emerged, life-like processes would have occurred naturally and spontaneously all over the planet. Given the right conditions and a spark of electricity, a 'primordial soup' of simple compounds can produce amino acids and other long molecules needed to build a cell. Amino acids can form in a wide range of conditions and are the building blocks of proteins, but to be able to do this, they must be concentrated. This can be achieved with heat, evaporation, freezing or even the help of special clay minerals. In truth, despite many experiments to recreate it, the early chemical history of life is still shrouded in mystery.

A little more illumination can be found when we look at the organic membranes that contain the concentrated ingredients. Without this boundary, none of the fundamental processes within would be possible. Membranes are often made of fatty molecules called lipids, which when added to water spread out like a film — you may have seen this happen in your kitchen sink when

LEFT: The Grand Prismatic Spring, Yellowstone National Park, USA. Life may have evolved in geothermal pools like this with the help of UV radiation from the Sun.

BELOW: With little land and a green-yellow hue, Earth before oxygen appeared quite different from the blue planet we are familiar with today.

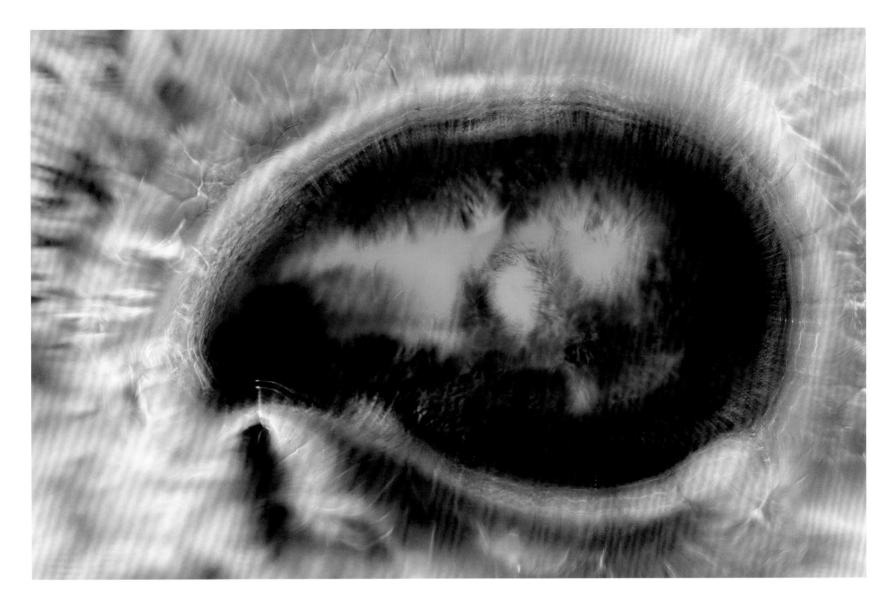

oily pans are soaking. This shiny layer forms because, on a molecular level, the lipids are aligning to position their water-attracting sections towards the water and the water-repelling section pointing away. When the mixture is agitated this surface film can fold in on itself, forming water-filled bubbles (or 'vesicles'), which could also contain amino acids and other helpful molecules. Like an organic test tube, these compounds could react without outside interference, as their own isolated system of chemistry. In most of these vesicles, the reactions would be unstable and peter out quickly, but a few were stable for longer. Eventually, some were able to take in compounds from outside and dispel anything unfavourable, and others must have divided as well, replication being a signature ability of true life.

Every cell requires information about how to build proteins, and successful replication depends on the offspring cell inheriting a copy of those instructions. Living cells today store this information on long strings of molecules such as DNA (deoxyribonucleic acid) and the more primitive RNA (ribonucleic acid). The point at which these little systems of chemistry began to replicate using RNA and compete for life was also the moment for the birth of the survival of the fittest (see page 30). For the first time in Earth's history, chemistry had become biology.

All life, no matter how different it may seem, has some genetic information and fundamental cell operations in common. This has a powerful and fascinating implication: every species that has lived and died in the last four billion years is not only related, but is also descended from one original ancestor. One of those primitive cells had a code that would be copied and modified by everything to come. This remarkable survivor has been called LUCA – the Last Universal Common Ancestor. By following the genetic trail back to its roots, we know that LUCA lived in hydrothermal vents using hydrogen as an energy source. LUCA was the deep-sea survivor that led to all life as we know it today, but its ancestors may, in fact, have originated on land. The last decade has seen fierce debate about where those earliest chemical systems for life began, a captivating mystery for future scientists to unravel.

TRACING THE ORIGINS OF LIFE

Identifying the earliest forms of life in the fossil record has proven very difficult, not least because single-celled organisms often resemble little more than blobs, and rot away quickly after death. Then there are the four billion years in which the fossils could be destroyed, either weathered away at the surface or melted deep in the Earth's crust. There is also no consensus about what the first lifeforms looked like or whether natural chemical processes could mimic its appearance. As such, contenders for the 'oldest life form' are always met with healthy scepticism by the scientific community; one person's round blob in a rock is another's first life.

Some of the oldest contenders are tiny branching filaments and tubes of iron oxide, found in particularly ancient Canadian rocks called the Nuvvuagittuq Belt. These structures may have been formed by primitive bacteria, adapted to live near hydrothermal vents under the sea. These vents are volcanically active zones of the seabed where the water can be super-heated to over 460°C and is laden with heavy metals and sulphur. It may seem like an underwater hell, but many microbes today thrive in these remarkable habitats. Indeed, whole ecosystems with complex food webs can develop around these underwater islands of volcanic fury, surviving on geothermal energy without a single ray of sunshine.

Other candidates for the earliest fossilised life are found in a 3.47-billion-year-old rock called the Apex Chert, in Western Australia. These tiny strings of blobs are so small that eight of them could fit along the width of a human hair. Like the older Nuvvuagittuq structures, they were likely produced in a hydrothermal setting, but this debate is very much ongoing.

The likelihood of finding the very earliest life, perfectly preserved, is vanishingly small – a microscopic needle in a planet-sized haystack. However, evidence for the earliest life can also be found preserved in the genomes of living species. By using the natural rate of change in genetic material through time, it is possible to calculate when groups appeared, even if there are no fossils. This method showed that life may have evolved 4.5 billion years ago, not long after the formation of Earth itself and the collision with Theia. It also means early life survived hundreds of millions of years of meteorite impacts throughout the Late Heavy Bombardment, and LUCA itself may have evolved during this onslaught.

From that single cell called LUCA arose countless generations of descendants, dividing and replicating at the expense of everything else. How did that one species become the millions on Earth today? The truth lies in imperfection. Every time a cell divides, the genetic code is copied and a clone of the parent is produced. However, very occasionally a mutation may occur that alters the genetic code. If this alteration is not fatal, it is passed on to the offspring and, before long, a population of mutants is living

RIGHT: Rock layers in East Pilbara, Australia, preserve red iron-rich sediments, rusted by the oxygen produced by Earth's earliest life forms over 3 billion years ago.

happily alongside the parent. If that mutation gives the offspring an advantage of some kind, then it is more likely to survive and reproduce in a given situation. For example, if a new variant can survive at higher temperatures, it could be said to have greater 'fitness' in a warmer-water environment. Likewise, if temperatures were cooler, then the original variant has the advantage, and would be fitter.

Natural selection is the process by which the fittest – or strongest – life form in a situation survives preferentially, and their genes are carried forwards. As environments and other survival pressures shift and change, so too does the relative fitness of each variant. Slowly these mutations can stack up, until they are genetically different enough to be called a distinct species. The basis of evolution is that more offspring are produced by a parent than can possibly survive and within those offspring variation creeps in. Natural selection determines which survive. The mutations themselves may be random, but the process of natural selection is brutally discriminate.

IDENTIFYING NEW SPECIES

Biodiversity is greater today than at any point in Earth's history and the number of species that are alive may be in the trillions. Despite enormous efforts over hundreds of years, the overwhelming majority of these species remain undescribed. Identifying a new species is no easy task, even if it is lucky enough to catch a scientist's eye. Some species can look almost identical even if they are genetically very different. Modern genome analysis has helped researchers look past appearances, but this is only possible when you have genetic material for the species. This kind of material is almost never preserved in fossils, which are often just scraps of bone or shell. We will never know quite how many species there were of *Tyrannosaurus*, for example, because we cannot sequence their genomes.

As well as genetic differences, an important definition of a species is whether it can produce fertile offspring. Tigers and lions can interbreed to make ligers (if the mother is a tiger) or tigons (lion mother), however, the offspring themselves cannot reproduce. This is a big problem for palaeontologists because they have no way of knowing which of the 'species' they have discovered were reproductively isolated like this. Also, males and females often look very different, so the sexes can sometimes be counted as two species. Only in a very rare few instances do we find fossils in the act of mating, which directly confirms they are sexes of the same species.

RIGHT: Complex and productive habitats, such as the Maldives coral reefs of the Indian Ocean, are biodiversity hotspots.

Another hurdle is that species can undergo massive changes in their appearance throughout their lives. Butterflies and caterpillars, frogs and tadpoles, human adults and babies – all of these look very different but are the same species. It is a confusing subject, and one that is debated more than anything else at palaeontology conferences.

So, why all the effort? The early days of describing and naming – a process called taxonomy – was driven by that odd human urge to collect. Victorian scientists – usually wealthy men – were curious and competitive and named every animal and plant within reach of their nets or guns. Modern taxonomy has become a lot more focused on the bigger picture and the way that ecosystems function. Having accurate information about biodiversity is a crucial baseline for conservationists who want to measure the ongoing health of a habitat.

Biodiversity through time is of great interest to palaeontologists because it reflects the conditions of Earth and can teach us about modern evolutionary processes. For example, when an unusual number of new species appears in the fossil record it can be an indication that the climate is favourable or, perhaps, that the ecosystem is getting more complex and productive. A drop in diversity, on the other hand, could signal a disastrous turn of events when a great many species are going extinct. While it is an uncomfortable thought, extinction is a very natural part of the history of life. There is a background rate of extinction that is quite normal in the geological timescale of millions of years. The balance of newly evolving species and the extinction of others is always in flux, but over the last 600 million years or more there has been a general increase in global diversity.

One important thing to bear in mind is that the oldest rocks also tend to be the rarest, as more time has passed for them to be naturally destroyed. Even when this is taken into account, it does appear that life has become progressively more diverse over time. The habitability of the planet is by far the most important factor affecting biodiversity. For life to flourish, habitats must have enough water, warmth, nutrients and, in most cases, sunlight. If those basic requirements are met, species naturally find balance in the ecosystem, and as time goes on more niches are found and exploited. Something like this process can be seen in miniature by leaving a boulder in a forest. At first, only bacteria and algae occupy the surface, and it is a relatively simple ecosystem of a handful of species. However, lichens and mosses soon follow and, before long, ferns are putting down roots. Finally, larger animals higher up the food chain move in, such as insects, until eventually hundreds of species may exist on this once-barren rock. Rather than simply arriving from elsewhere, though, each species evolves to fill the vacant niches and build an ecosystem. Natural increases in ecosystem complexity rely on the stability of climate and nutrients, but on a global scale these things are constantly changing.

RIGHT: The tadpoles and adults of strawberry dart frogs look very different, so their fossils would give little clue that they are the same species.

NEXT PAGE: Iceland began forming 16 million years ago as tectonic plates under the Atlantic Ocean drifted apart and volcanoes brought new rock to the surface.

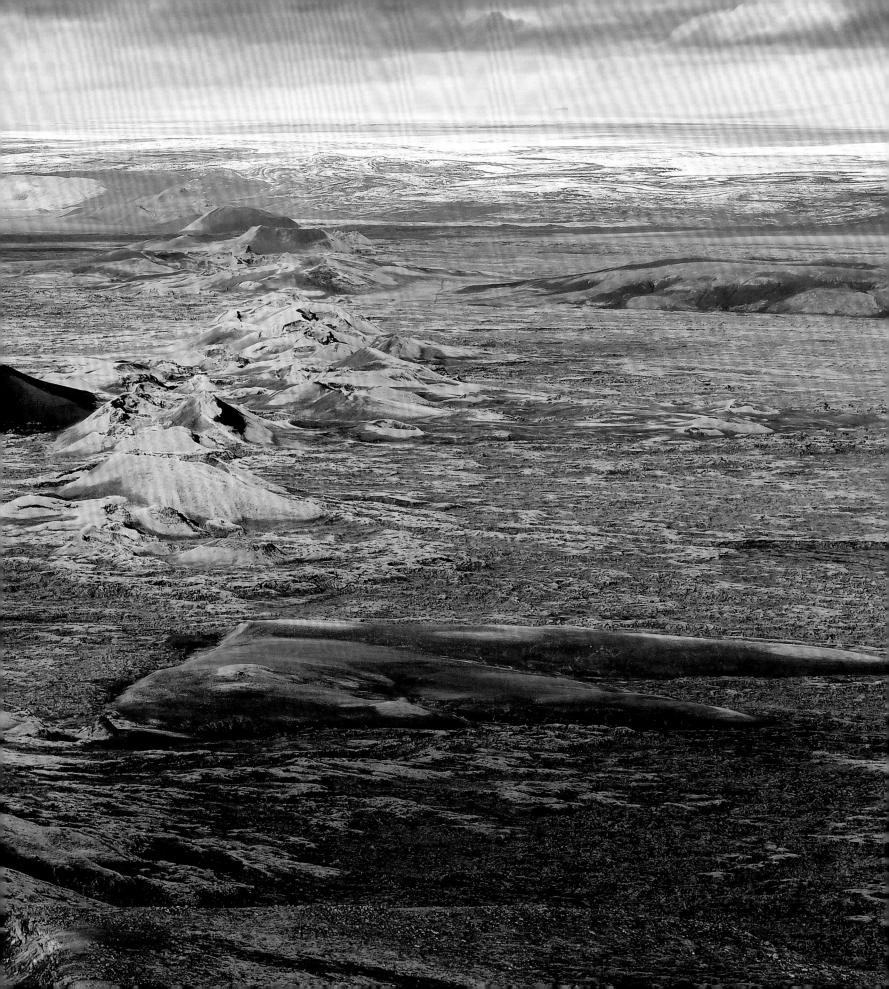

OUR SHIFTING WORLD

Over hundreds of millions of years, the configuration of Earth's continents has shifted beyond recognition. The rocky rafts that we all live on (the lithosphere) can be moving at up to 16 centimetres a year, scraping past each other, drifting apart or even colliding head on. The engine driving this tectonic dance of landmasses is heat from the molten mantle below. Huge convection currents of magma carry intense heat from the depths up towards the surface, before arching over and dragging the solid surface plates with them. As the magma cools, it begins to sink back down towards the Earth's core, before reheating at depth and circulating up and around all over again. This magma can be as little as 5 kilometres below the surface, and for every kilometre you dig down the temperature will increase by up to 25–30°C. The deepest mines in the world have to contend with this heat to operate but, remarkably, even at depths of 3 kilometres or more, bacterial life can eke out an existence.

Plate tectonics has – since this movement began around 3.4 billion years ago – created and destroyed habitats like nothing else and this activity continues to this day. It can create mountains, fuel volcanic eruptions, earthquakes, sea-level changes and even affect the climate and oceanic currents. It is a climate that has had a particularly powerful effect on life, as it governs the habitability of the planet. Volcanoes can release large quantities of gas when they erupt, such as sulphur dioxide, which can cool the atmosphere. However, in the longer term another volcanic gas, carbon dioxide, is also released, which has a warming effect. Today, volcanoes emit around 300 million tonnes of carbon dioxide a year and, until humans arrived, were the chief source of carbon dioxide in the Earth's atmosphere. Warmer global temperatures aren't necessarily a bad thing, and they can open up habitats in higher latitudes, such as when dinosaurs roamed the forests of Antarctica in the Cretaceous period. Rather, it is during a rapid change in temperature, when species cannot adapt quickly enough, that mass extinctions occur.

LIFE AND EARTH – A FINE BALANCE

The other side of the coin is that life processes can also dramatically affect the Earth itself. Fungi, for example, can chemically dissolve minerals to feed themselves, and plant roots do a fine job of breaking up solid rock to anchor themselves in the soil. On a larger scale, nutrients like phosphorus, which would otherwise be trapped beneath the ground, are brought to the surface by these biological processes. As well as speeding up the breakdown of rocks, animals and plants can create them; take a walk along a city street and it does not take long to find fossilised shells in the building stones. Some rocks, such as chalk and coal, are made up entirely of fossilised remains – chalk is composed of microscopic skeletons, while coal is formed from ancient plant matter.

Life and Earth processes are intertwined in a great many ways, but one of the most important is the balance of gases in the atmosphere. Even the earliest single-celled life, four billion years ago, would have exerted its influence on the skies above. These organisms released the greenhouse gas methane as a waste product, so the first signal of life's existence on Earth may have been a yellow haze that warmed the ancient air. Today, similar organisms live at hydrothermal vents in the ocean, geothermal springs on land, and even deep beneath the Earth's surface in mantle-heated groundwaters. These simple and primitive lifeforms were fed by a soup of volcanic

BELOW: Gases released by volcanic activity have always had a profound influence on the Earth's climate.

nutrients dissolved in water and would have been restricted to habitats that could provide it. That all changed when one cell, among countless trillions, evolved a chemical pathway that allowed it to use carbon dioxide and sunlight to make energy. This innovation was monumental, especially amongst the throngs of competitors all trying to gain access to limited nutrients and space. By exploiting a new source of energy, this new breed of organism would have had a huge survival advantage. Successive generations would have passed on their genes, mutations would creep in, and natural selection would continue to drive the evolution of even better sun-fuelled cells. As the process became more efficient, the descendants could gradually cut their ties to volcanic energy and spread to new habitats.

Despite being microscopic as individuals, we do find colonies of these bacteria preserved in the fossil record, dating to perhaps over 3.4 billion years old. These fossil structures are called stromatolites and are made up of thin layers of sediment piled up to resemble lumpy mounds. Today, photosynthetic bacteria that create similar structures produce an adhesive gloop to cement themselves to sunlit surfaces. Over time, particles in the water become attached to this adhesive and accumulate there, blocking the light and forcing the bacteria to migrate upwards. As this happens, they leave a barcode of layered sediment below them, so a colony can produce mounds that exceed a metre in height.

Early photosynthetic bacteria were free to conquer the Earth and did so for around 600 million years before a major innovation fundamentally changed the course of history. Around 2.8 billion years ago, a new type of photosynthesis evolved, which released oxygen as a waste product. Until this point oxygen made up less than 1 per cent of the atmosphere, compared to 21 per cent today. The innovators that were producing this oxygen were cyanobacteria, and they had a unique advantage. As they produced waste oxygen, this effectively poisoned the more primitive (anaerobic) bacteria. Until this point, the seas had been a light green colour, tinted by forms of iron that had reacted with sulphur or chlorine. As oxygen crept up, the dissolved iron oxidised, turning the seas a deep brick red. Within a few hundred million years, these oxides began falling out of suspension, carrying huge amounts of iron and oxygen that built up on the seafloor; so much, in fact, that almost two-thirds of all the iron we use today comes from these deposits. It is wonderful to think that the iron and steel objects around you right now came from a great rusting of the oceans two and half billion years ago, during a microbial world war. The iron in the water and anything else that could oxidise formed huge chemical sponges for the oxygen being created by cyanobacteria. As oxygen was created, it immediately reacted, and this continued until the reactions slowed down and free oxygen began

RIGHT:
The stromatolites of Shark Bay, Australia, are created by cyanobacterial colonies.

to finally escape and build up in the atmosphere. It appeared as if the cyanobacteria had won, and the anaerobes were relegated to pockets of stagnant, low-oxygen habitats like the deep oceans, a shadow of their former glory.

The cost of cyanobacterial success was that their oxygen production was fundamentally changing the Earth's climate. Methane, one of the most powerful greenhouse gases in the early atmosphere, was reacting with their waste oxygen to produce water and carbon dioxide. Meanwhile, carbon dioxide was being consumed by the very same cyanobacteria to make carbohydrates. Greenhouse gases were being removed on two fronts, and as that happened global temperatures began to drop. Water began to freeze and the white ice reflected more and more of the Sun's heat away from Earth. Temperatures spiralled downwards, then by approximately 2.4 billion years ago, the Earth had entered an ice age.

It is worth taking a moment to imagine what this world looked like. The Moon's orbit has been steadily widening by about 3.8 centimetres per year, and from the moment it formed, its gravitational influence has slowed the Earth's rotation. So, about 2.4 billion years ago, the Moon would have looked enormous in the night sky, exerting a massive tidal influence on the oceans. At that time, an Earth day lasted only 17 hours. At the height of the oxygenation event, the sea would have been blood red with iron and a methane haze would linger yellow in the sky. It was an utterly bizarre vision of Earth.

This would have been a turbulent time for cyanobacteria, which relied on carbon dioxide to live. Throughout the next 200 million years, ice ages came and went as pulses of cyano-bacterial boom and bust caused temperatures to rise and fall. Around 2.1 billion years ago, the glaciations ended and a more hospitable planet was revealed. Oxygen high up in the atmosphere was being bombarded with ultraviolet light rays from the Sun, creating a thin ozone layer. This, in turn, blocked a lot of UV radiation, protecting organisms from genetic damage and allowing them to get larger and more complex. The stage was set for the next great leap of evolution.

A NEW GENERATION

Since LUCA, the Earth had been ruled by a group of single-celled organisms collectively known as prokaryotes. These were relatively small and simple, with a ring of DNA, which floated freely around the cell. Despite the lack of complexity, there were many species and all played their parts in microscopic food webs. There were prokaryotes that used the Sun's energy to

LEFT: Cyanobacteria, like this *Rivularia* colony, produced the oxygen needed for animals to evolve.

make food and predatory prokaryotes that fed on them. At some stage amidst the ice age, a prokaryote was engulfed by another cell but, rather than digest the prokaryote, the cell kept it alive. Whether this was intentional or a fluke occurrence is still unclear, but the act itself was a significant leap. This cell and its tiny internal counterpart were the first of a new lineage, the eukaryotes.

Today, the eukaryotes include every living organism you can see and many more, from yeast to yaks. The cells are larger and more complex than prokaryotes and package their DNA in a membrane called the nucleus. Eukaryotic cells also contain membrane-bound organelles that perform functions for the cell. Some of these have their own independently replicating DNA, supporting the idea that they were once prokaryotes engulfed by the cell and then used for its own benefit. Within the host cell the prokaryote released energy that the host could use.

BELOW:

Chloroplasts, which green plants such as water lilies use for photosynthesis, were originally free-living cyanobacteria.

In return, the host provided nutrients, a home, and a vessel for tasks the prokaryote could never achieve on its own. The process of engulfment for mutual benefit like this is called endosymbiosis, and it may have occurred several times throughout the story of life.

This initially happened perhaps around 2.4 billion years ago, when fungus-like cells are found in the fossil record for the first time. In this case, the captive prokaryotes became the organelles we know as mitochondria. Later, around 1.6 billion years ago, cyanobacteria were engulfed and became chloroplasts, helping the plant lineage to gain energy from the Sun. Throughout evolutionary history, collaborations like these have happened time and time again between species, functioning together to improve survival.

It is worth reiterating just how incredible the prokaryotes are. Throughout their four-billion-year tenure, they have become the most diverse and numerous organisms on the planet. In a litre of lake mud there may be 460 billion prokaryotic cells living their lives. Despite being ten times smaller than eukaryotes, and with perhaps 1,000 times less DNA, they have conquered almost every habitat on Earth. The spirit of collaboration is still very much in effect, and prokaryotes routinely help more complex organisms perform biochemical reactions. In your body there are roughly the same number of prokaryotes as human cells, and at this very second trillions of them are busy helping you digest food.

While undoubtedly successful, prokaryotes are limited in the way they reproduce. Every time they divide the genetic material is copied and you are left with two identical clones. This is called asexual reproduction and is great for rapidly increasing population numbers, which effectively double with every generation. However, with every cell being genetically similar there is not much for natural selection to act on. Evolution in prokaryotes is therefore relatively slow and relies on mutations creeping in, or very occasionally new genes being shared between species. Many eukaryotes, on the other hand, are capable of something quite different – sexual reproduction. This means that the DNA of two parents is mixed and recombined in the offspring. Because of this the next generation is similar but not identical to the parents, which means they have a larger palette of genetic diversity. Every one of the offspring is genetically unique, and therefore more or less likely to survive.

Sexual reproduction creates a variation on which natural selection can act more quickly. For example, if a green stick insect, living in a forest, reproduced asexually the population could grow very quickly, but all the offspring would be identical. If there was a dry spell and the leaves browned, all of the green clones of the parent would be vulnerable to predators and be picked off quickly. However, if two stick insects reproduced sexually, there would be a variety of young born. Some might be marginally browner than others, and when the dry spell comes they are slightly less likely to be eaten. The downside to greater offspring variety is that sexual reproduction requires two individuals finding each other to create offspring. So not only is the effort required to reproduce like this greater, but it also produces half as many offspring per parent. Despite this, it is the most common method of reproduction in animals, plants and fungi today, demonstrating just how powerful and beneficial it is.

At around the same time as sexual reproduction was evolving, the eukaryotes were experimenting with a very ancient survival strategy. Individual cells must be self-sufficient, but some prokaryotes, like cyanobacteria, had been forming colonies for more than a billion years. Together they were more resilient, more efficient, and could achieve a division of labour that helped them all. Reaching that point was fraught with obstacles, not least the inbuilt desire for each and every cell to survive and spread its own genes.

At some point, perhaps 1.8–2 billion years ago, eukaryotes managed to make that leap to true multicellularity for the first time. They achieved the harmonious organisation of cells cemented together performing different jobs. The cells had to communicate, coordinate, and even die at a predetermined

RIGHT: These *Heliconius* butterfly eggs are near-identical in appearance, but the genetics of each egg is completely unique.

moment for the greater good of the larger entity. Complex multicellularity would eventually occur in six eukaryote lineages, including various algae, land plants, some fungi, and of course animals, testament to the advantages of cells giving up their individuality to form something greater than the sum of their parts. However, a massive setback was on the horizon for eukaryotes.

A PARTING OF THE WAYS

Plate tectonics was still in its infancy and at around 1.8 billion years ago, for whatever reason, came to an abrupt halt. Without volcanic activity delivering nutrients to the surface, organisms that photosynthesise could no longer produce oxygen as they had been. Without tectonic activity the oceans were left poor in nutrients and oxygen, and a choking milky soup of sulphides developed. For the first time since the great rust, oxygen levels were low enough that the prokaryotes could rise from the deep oceans to reclaim the shallows. For the next billion years, plate tectonics would lie in stasis, the climate would remain unchanged, and waters would remain starved of nutrients and oxygen.

For the Earth itself, this was a remarkably uneventful time – even earning the name 'The Boring Billion'. This is, perhaps, a little unfair when considering the eukaryotes' story. With oxygen at around 1–2 per cent of modern levels, the eukaryotes were well and truly the underdogs of the time, clawing out a meagre existence compared to their prokaryote elders. In fact, this adversity may have pushed the marginalised eukaryotes to step up their game. Around 1.4 billion years ago, they began to diversify, becoming more complex and widespread. The plant line may have split away from the ancestors of animals and fungi at this time, closely followed by animals and fungi parting ways themselves, off on their own evolutionary journeys. It should be noted that these are educated estimates and we have yet to find irrefutable fossil evidence for these dates. The likelihood of us finding fossils from the exact moment that two groups diverged is vanishingly small, so instead palaeontologists apply a process called the molecular clock.

Using the predictable rate of genetic mutations through time, it is possible to back-calculate how old a lineage is, even without fossils. Any you do find can be used to improve the calculation and fine-tune the date, and this happens all the time with new discoveries every year.

By the end of the not-entirely-boring billion, Earth's plate tectonics were back in action. An early supercontinent known as Rodinia was beginning to fragment, and through the fractures,

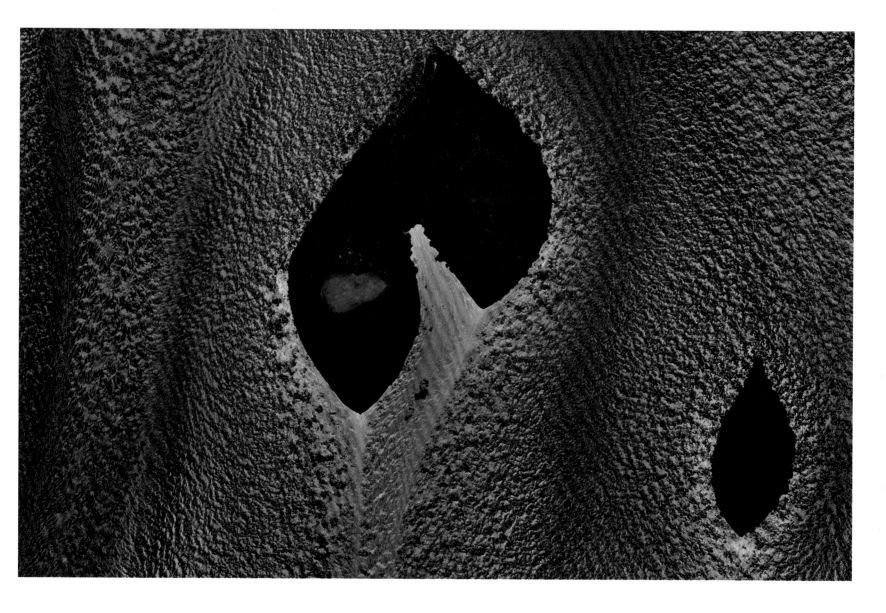

lava was gushing to the surface. Volcanic rock is full of nutrients and, as it weathered away, these nutrients slowly replenished the long-barren oceans.

For billions of years, simple prokaryotes had occupied the volcanic landscape, hugging moist rocks near rivers and lakes. Then, around 1 billion years ago, they were joined by the first eukaryote pioneers, primitive fungi that clung to existence with the prokaryotes as films and crusts on moist rocky surfaces. These organisms may have accelerated weathering of the volcanic rock, helping to release nutrients into waterways. As this fertiliser made its way to the sea, it could have fed photosynthesising organisms, which began to produce oxygen as a waste gas. Higher oxygen was the accelerant needed to fuel the increased size and complexity of eukaryotes, and before long, at around 800 million years ago, some of the first animals — sponges — appeared.

ABOVE: The 2021 Cumbre Vieja volcanic eruption in La Palma. Volcanic outpourings of lava like this bring exotic chemistry to the Earth's surface, the essential nutrients for life.

Sponges are multicellular but, unlike most of the animals we know today, they do not have distinct organs, or blood or nervous systems. Instead, they are more of an organised mass of cells and hard structures called spicules, all glued together with collagen. In order to feed, they pump water through pores and canals, filtering it along the way. They were huge advances from the single-celled eukaryotes of old, but primitive and simple by today's animal standards.

Things seemed to be looking up for eukaryotes, but this oxygen boost came with a huge price tag. As well as releasing oxygen, the photosynthesisers were using up carbon dioxide and causing it to deplete rapidly in the atmosphere. On top of this, the newly erupted volcanic rock was chemically reacting with carbon dioxide, locking it away as carbonate. Without this greenhouse gas, the Earth was plunged into an incredible deep freeze. This phase is often called 'Snowball Earth', because during this time two pulses of glaciation transformed the planet into an ice world. While there was a brief respite between them, these two glaciations were particularly intense. At the poles, it may have reached -130°C, and perhaps 0 to -43°C at the Equator.

Beneath the ice, the oceans were in stasis, slowed to an almost complete standstill by the darkness and bitter cold spells. Nevertheless, the sponges soldiered on undeterred, even though oxygen levels were often incredibly low. At the Equator, the ice may have been thin enough to let some light through, allowing cyanobacteria to photosynthesise and produce great swings up and down of oxygen for the sponges below. Sponges would also be filtering out these cyanobacteria as a food source, which begs an important question: if sponges needed oxygen to live, how could they also feed on the organisms making that oxygen for them? Some scientists believe that by consuming nutrients and huge numbers of the tiny cyanobacteria, sponges may have forced the evolution of larger and more efficient photosynthesisers, such as algae. Sponges were silently engineering a higher-oxygen planet.

As we've seen, one of Earth's great thermostats is the chemical reaction of carbon dioxide with exposed rock. More exposed volcanic rock helps remove more carbon dioxide, a greenhouse gas. However, during the glaciations of Snowball Earth, a lot of this rock was covered in ice, preventing these reactions. Of course, erupting volcanoes also release carbon dioxide, and this may have been what finally warmed the planet. Slowly but surely, the balance was tipped, carbon dioxide began to accumulate in the atmosphere and the temperature rose once more. The ice began to melt and, by 635 million years ago, it had thawed, revealing a planet primed with animal life and the oxygen it needed to thrive. This new world was the Ediacaran; rain fell again, rivers ran and the warmed Earth was once more a blue planet.

LEFT: Sponges, like these *Anoxycalyx* in Antarctica, are some of the most primitive animals on Earth, and incredibly hardy.

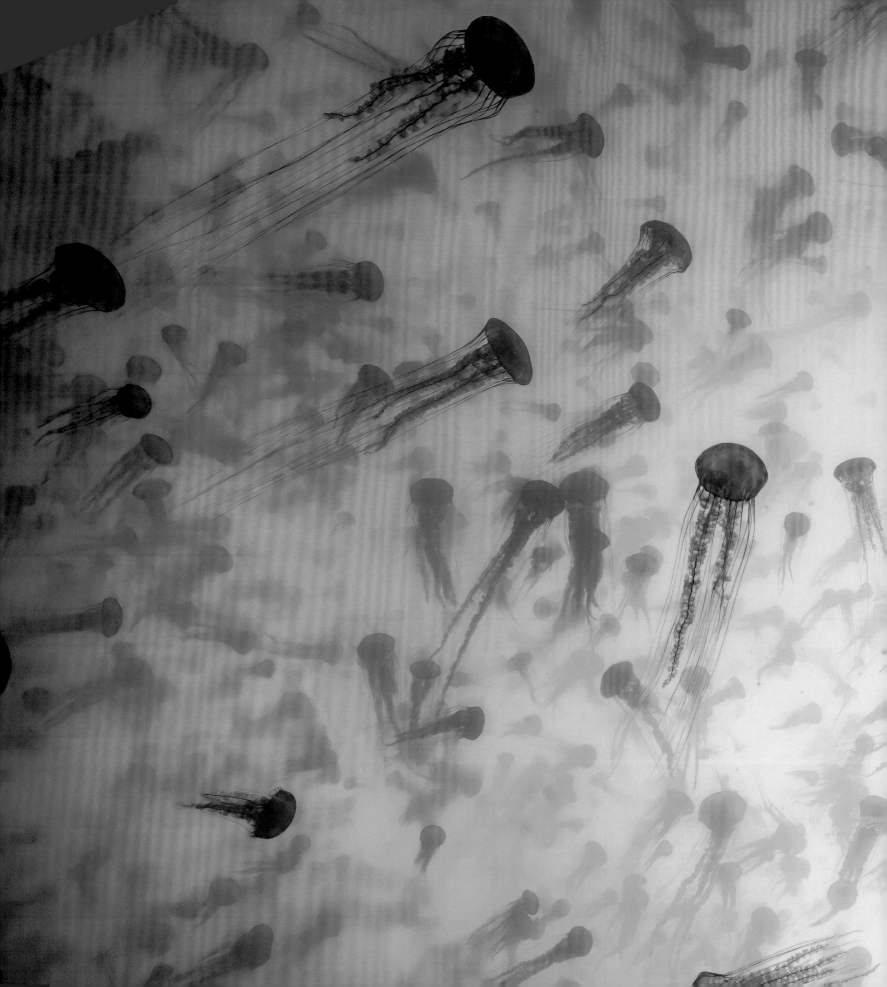

THE FIRST FRONTIER

In the early Ediacaran period, about 609 million years ago, the fossil record teases us with tiny glimpses of diverse animal life. Perhaps the most astonishing of these discoveries are minute structures from South China, found in their thousands in layers of rock called the Doushantuo Formation. These microscopic fossils are so small that they have to be picked out individually with a paintbrush and studied using high-powered x-rays. Despite being hundreds of millions of years old, the detail of the individual cells and maybe even the yolks are preserved in three dimensions. For a while, these were thought to be fossilised embryos and were celebrated as a missing link between simple eukaryotes and animals in the depths of the Ediacaran. Recently, however, it has been suggested that the fossils are likely to be more primitive eukaryotes. When times are tough, single-celled protists — a group of relatively simple animals that include amoeba and *Paramecium* — can hunker down in specially constructed capsules, called cysts, until conditions improve. The strong resemblance of these kinds of cysts with the Doushantuo fossils may mean they are not true animals after all.

THE FIRST ANIMALS

Things become much clearer later in the Ediacaran, from around 575 million years ago, when numerous complex organisms appear in the fossil record. Before their discovery scientists

PREVIOUS PAGE: Jellyfish and their relatives were amongst the first complex animals to evolve, igniting an evolutionary arms race that rages to this day.

RIGHT: *Dickinsonia* could grow to over a metre long and was symmetrical along the middle of its body.

believed that the first four billion years of Earth's history were quite dull and uneventful. However, this disinterest began to change in the 1950s after schoolchildren discovered an Ediacaran animal called Charnia in Leicestershire, England. This was enough to convince palaeontologists that the story of life during the Precambrian was much more interesting than previously assumed.

The oldest Ediacaran animals come from Newfoundland, in Canada, but similar fossils are found in younger rocks in Australia, Namibia, Russia and the United Kingdom. These creatures became very diverse, with over 200 known species ranging from around 1 centimetre to 1.4 metres in size. They were all soft-bodied, but they left various imprints that resemble long fern leaves, discs, bags and odd circular quilts. Because of their bizarre and unique appearance, scientists had debated for decades about what modern animals these organisms could be related to. Until very recently, it was thought they could all just be related to each other, forming their own taxonomic group entirely, which eventually died out – a failed experiment in evolution. However, thanks to chemical analysis, we now know that at least some of these Ediacaran forms were, indeed, true animals, and their diversity suggests they formed the first major sea-floor communities. Some species may have absorbed nutrients passively from the water or filtered it out of suspension. Others may even have moved around, grazing on mats of algae and bacteria that grew in the shallows.

Movement and the ability to travel is something we now take for granted, but many species of the Ediacaran were fixed to one spot. They were at the mercy of whatever nutritional titbit floated their way, unlike those that could actively seek out greener pastures. The advantages of travel meant that, before long, a familiar body plan had evolved, which made this more efficient. Most animals have either bilateral or radial symmetry; that is, their body is built in two mirror-image halves (bilateral), or around a central point like an umbrella (radial). Radial symmetry can be found today in things like starfish, jellyfish and sea urchins, and bilateral symmetry in every vertebrate you can think of, including yourself. Just like those early Ediacaran movers, we have a right side of our bodies, which is a mirror image of the left. This makes the body much more streamlined and easier to move with less energy in one straight direction. Some early movers even went one step further and developed a head end, which could lead the direction of movement. Again, we take it for granted that we have a head, especially one with such a convenient arrangement of mouth, nose, ears and eyes. One particular Ediacaran species of note is the tiny, 3-centimetre-long *Spriggina*, which vaguely resembles a trilobite but may be more closely related to annelid worms, no one really knows. Most significantly, *Spriggina* has a distinct head end, which scientists believe may have had sensory organs, perhaps even eyes. Seeing, smelling or tasting food, before they ingested it, was an important step for these creatures and the animals that followed. Whether animals like *Spriggina*

RIGHT: The Indonesian fire urchin, like all echinoderms, begins life with bilateral symmetry before becoming radial as it matures.

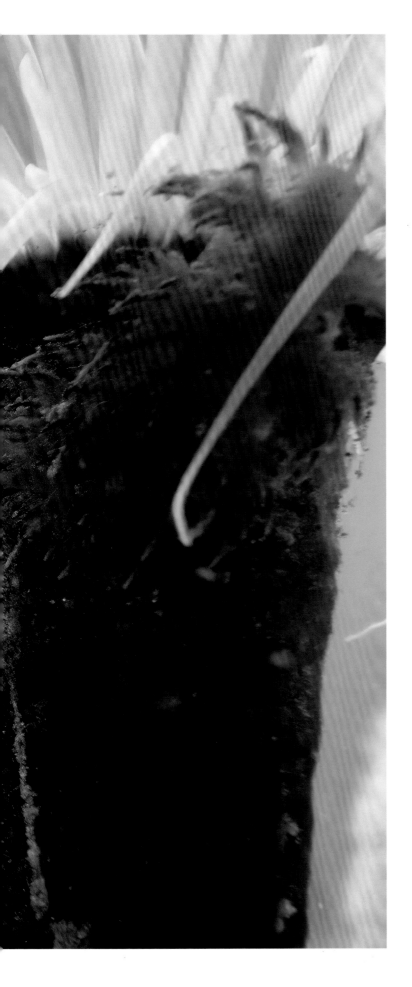

were becoming more discerning herbivores at this point or something more sinister, we will never know. However, the fact remains that in the 30 million years of Ediacaran diversification, a smorgasbord of defenceless fresh meat was on the table. The age of innocence for animal life ended when they started eating each other, and the escalating war between predator and prey began.

SURVIVAL OF THE STRONGEST

Predation was not a new concept; microbes had been doing it for billions of years and for good reason. Why go to the trouble of producing complex molecules when you can simply consume them, premade by another organism? This escalation forced the hand of potential prey, who could no longer afford to be squishy and defenceless. At around 550 million years ago, a large number of unrelated animals began to reinforce their bodies with hard minerals, creating tough skeletons and armour. In under 25 million years, the soft-bodied Ediacaran animals vanished, replaced by an army of aquatic warriors.

Until this point animals had been relatively simple, often living happily fixed to the sea floor or grazing slowly on the surface of the water. However, many groups had been experimenting with mineralisation, adding hard crystalline materials to their bodies.

LEFT: The rainbow nudibranch is a specialist predator of tube anemones that anchor themselves to the sea floor.

Sponges had been doing this for some time, using pointy structures known as spicules to help support their bodies. By creating these hard silica elements and meshing them together with their cells, sponges had created a primitive and makeshift skeleton. More-advanced animals are faced with a choice: either they keep a soft deformable body, capable of things that would make a contortionist jealous, or they create a skeleton. A skeleton is a hard framework for muscles to attach to, which add structure to the body. This can be very useful if the animal needs leverage and strength to move, but it can limit movement as well. Often there is a trade-off between the two states depending on the animals' lifestyles. Octopuses, for example, are incredibly flexible and have very few hard parts in the body. The cost of this flexibility is that they are soft, fleshy and vulnerable to predators. At some point during the latest Ediacaran, around 560 million years ago, some animals began toughening their external body surface, fashioning the first exoskeletons. This may have first evolved to protect against mechanical damage from the environment and maintain a rigid home within which the animal could peacefully live and feed.

The rise of predation, at around the same time, would drive a whole suite of animals to improve their defences. Even a little toughening of the body surface would have helped, while the undefended species made easy snacks for predators. The first exoskeletons were stiff, cone-like tubes, which served as refuges to protect the soft parts of the animal. This is still a very successful strategy today for things like snails and corals, which can quickly retreat into their homes. Despite their tough shells, snails are still vulnerable when moving and must stop what they're doing and hunker down when a predator is near. The most effective armour would be a solid, continuous shield around the entire body, but this would prevent the animal from bending. The compromise between range of motion and protection is a familiar problem for human soldiers. Throughout history all manner of body armours have been tried, but most rely on a similar principle – by covering the body in smaller overlapping plates, the animal can bend freely while maintaining a relatively tough exterior.

No other group has adopted this strategy quite like the arthropods, a group that includes creatures as diverse as insects, spiders, scorpions, centipedes and crustaceans. Arthropods embraced external armour for protection, and it was so successful that there are over 10 million species in existence today. Evolving hard armour allowed all manner of creatures to survive on the sea floor, but this was a fiercely competitive environment. As well as armour, many animals were evolving hard parts capable of biting and grinding up food more efficiently; food that included defenceless soft-bodied animals. For them, the choice was stark – stay and fight, or flee. But where to go?

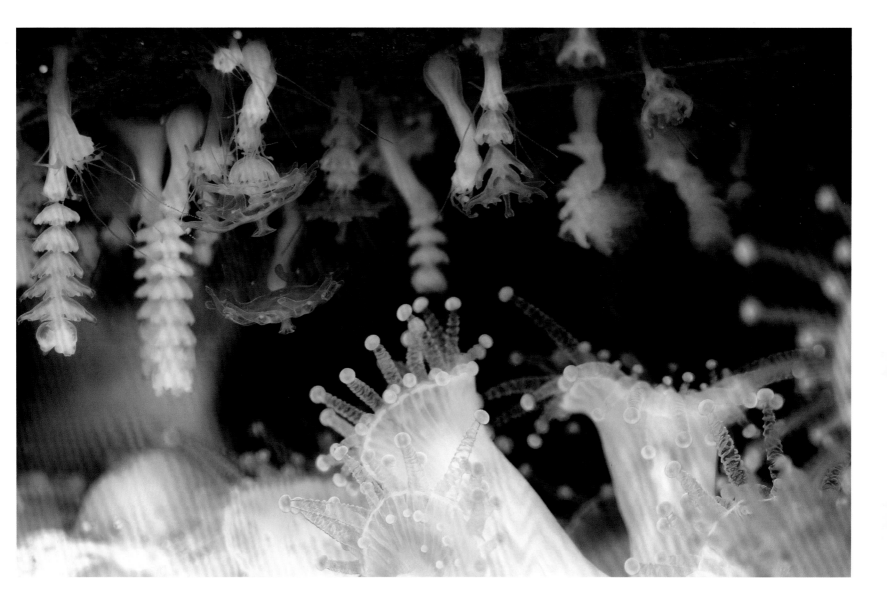

ONWARDS AND UPWARDS

For billions of years simple microscopic lifeforms had solved this problem and were truly liberated from the competition below. Instead of cementing themselves to a hard surface along the sea floor they allowed themselves to drift and soar upwards into the water, free to roam wherever the currents took them. These creatures, known as plankton, have little say over the direction in which they will travel, but in turn they have access to nutrients, light and vast amounts of space. The first animals to explore life in the open ocean were jellyfish, members of a particular ancient group called the cnidarians.

The first cnidarians had stalked bodies with finger-like tentacles at the top, resembling modern sea anemones. The modern jellyfish we think of are disc- or bell-shaped swimmers, but in fact this is just one stage of their life cycle. In early life jellyfish live just as their ancient ancestors

ABOVE: Adult jellyfish, known as medusoids, were the first large animals to conquer the open oceans.

did, as polyps anchored to the sea floor with tentacles pointing upwards to capture food. However, as this polyp matures it undergoes a strange transformation. Slowly, it becomes ridged, then the tentacles begin to shrink. Before long, the ridges become distinct segments over the length of the body, and at a critical point these segments break away entirely to make a new jellyfish. In a remarkable process known as strobilation, jellyfish babies pop off one by one and swim away. In the open water, they feed, grow into the familiar-looking adults and eventually spawn. There are usually male and female jellyfish, which are triggered to spawn *en masse* at dusk or dawn. The fertilised eggs develop into larvae, which eventually settle on a solid surface, each one transforming into another polyp.

Jellyfish were among the first animals to conquer the open oceans, breaking free from the sea floor and escaping into the big blue. But there was an alternative escape route from the hustle and bustle of the seabed: digging downwards. For millions of years, burrows were simple vertical or u-shaped tubes, which probably only served as retreats for the animals that made them. However, at the start of the Cambrian period, around 538 million years ago, animals began excavating much more complex burrows. It was a time of innovation of both behaviour and body shapes. Some early Cambrian

RIGHT: Strings of jellyfish medusoids, ready to break away and be free in the ocean, hang above the deadly tentacles of sea anemones below.

animals were beginning to live for extended periods below the surface, creating complex burrows as they mined through nutrient-rich layers of sediment. We will likely never know what kinds of animals produced these burrows, perhaps some worm-like creatures, but whatever their affinity, they represent a massive step forward for animal life, exploring every dimension of their aquatic world. Diversification was accelerating, and ecosystems were becoming more and more complex.

INNOVATIONS IN THE CAMBRIAN

Terms like 'Ediacaran' and 'Cambrian' may sound confusing, but it is the language of geologists and a standardised way to compare rocks around the world. By using this system, a researcher from Australia, for example, can study a layer of sandstone, knowing it is the same age as a limestone in the USA. From that information they can begin to reconstruct the environments of each place at specific times and begin to reconstruct the story of the Earth. By examining whether rocks around the world were laid down in the ocean or on land, maps of the ancient continents can even be drawn up in a huge international and collaborative undertaking by geologists. From these kinds of studies we know that in the Cambrian period there were

RIGHT: Brown noddies perch on layers of volcanic ash, also preserved as solid rock on Isabela Island, Galápagos.

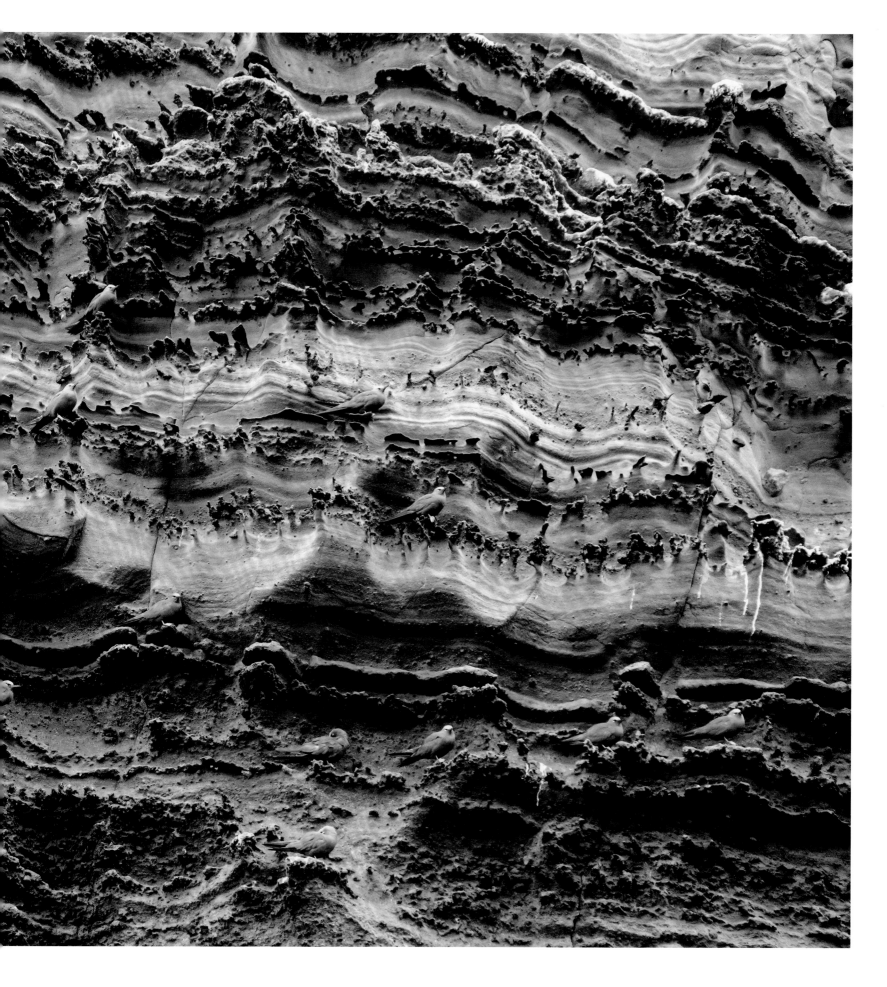

around four large continents that were mainly located in the southern hemisphere, including the supercontinent of Gondwana at the South Pole. Large areas of these continents were flooded, creating shallow seas, and while there may have been ice at the poles occasionally, temperatures were generally warmer than today.

After 3.5 billion years of relatively simple creatures, a burst of evolutionary innovation was setting loose new body plans and behaviours. In just 10–20 million years – a geological blink of an eye – almost every major group of animals had evolved. This momentous event was so abrupt and so significant in the story of life that it is known as the 'Cambrian explosion'. The cause of this diversification event is still unclear, but it may be related to changes in the ocean's chemistry, particularly increased oxygen levels. Algal blooms, at around the same time, would have enriched the water with oxygen during photosynthesis. This in turn would have provided the metabolic fuel necessary for animals to get bigger and live more energetic lifestyles. It is also possible that an abundance of calcium was eroding from rocks on the land and washing down rivers and out to sea. This mineral-rich water may have made it easier for animals living in it to build hard tissues like shells and skeletons, which were crucial to their survival amongst the crowd of new species. The Cambrian was also a time of relative warmth, which caused sea levels to rise, increasing the habitat available for life to thrive.

One of the first glimpses we get of the Cambrian explosion in the fossil record is of the Chengjiang animals of Yunnan Province, China. Around 200 species have been described from this locality, but what really makes these fossils so special is the detail of their preservation. Despite being 518 million years old, the rocks contain the remains of even very fragile organisms like algae, worms and sea anemones. In fact, the majority of the Chengjiang animals have no hard parts, making their preservation even more remarkable. Features that would have normally rotted away can be seen as thin rust-coloured films in the muddy layers. These include very delicate structures such as gills, guts, muscles, nerves and even eyes. The first fossil brains can also be found at Chengjiang, including the 4-centimetre-long arthropod relative *Fuxianhuia*. Despite its size, this strange little creature, which resembled a sea monkey, also preserved a heart and blood vessels, which represent the oldest cardiovascular system in the fossil record. The quality and quantity of fossils in these rocks are both truly exceptional and offer a snapshot of a bustling sea floor community during the Cambrian explosion.

Representatives of all kinds of animals are found at this site, with a huge array of body shapes and sizes. In total, around 17 animal groups have been found there, including some that look

RIGHT: Soft animals, such as sea pens, are not normally preserved, and so are underrepresented in the fossil record.

nothing like we have today, evolutionary experiments that for whatever reason did not stand the test of time. Perhaps most intriguing of all is that amongst these weird and wonderful creatures we also find members of our own group, the chordates. Today, the chordates include familiar animals such as fishes, amphibians, reptiles, mammals and birds, but the oldest members of the group are much more humble. One of these was the tiny Chengjiang creature *Myllokunmingia*: no larger than the end of your thumb, unprotected, and at first glance not particularly remarkable. In life, it would have resembled a tiny swimming worm, but this animal was set apart by a stiffened rod of cartilage, called a notochord, running along the length of its slender body. By contracting muscles on one side of the tail, as the other side relaxes, a curve is formed and if each side is contracted consecutively, a crude flapping motion is achieved. The first swim would not have looked pretty, but it would have allowed a quick burst of speed to escape predators. As it was refined by evolution, swimming would become more elegant and efficient, allowing these animals to explore the water column. *Myllokunmingia* had a distinct flattened area or fin around the tail end, which would have increased the thrust produced by each stroke, suggesting it spent most of its time swimming above the sea floor. This miniscule creature also appears to have had an internal skeleton and at the head end there is a skull, making it the first fish and one of our earliest ancestors.

FOSSILS – A WINDOW INTO THE PAST

The story of our ancestors and life itself can only be told because of centuries of scientific detective work. Studying modern life is a relatively easy job because the creatures are still around to observe and study directly. Palaeontologists, on the other hand, only get glimpses of these extinct creatures in their long-dead remains, trackways and the echoes of biology in their living relatives.

The fossil record is like a book with most of the pages ripped out, so any surviving information is precious. The likelihood of an animal becoming fossilised in the first place is miniscule, and it is even less likely that it will draw the attention of a curious researcher. The number of fossils that must have been exposed at the surface only to be weathered away is an eyewatering thought. Nevertheless, over a timescale of hundreds of millions of years the odds, like the rocks, have stacked up in our favour, and museum cabinets are full of fossils. The fact remains that the most likely parts of an animal to fossilise are the hard components, such as bone, teeth or exoskeletons. This certainly helps explain the lack of fossils before the Cambrian. For

RIGHT: The hard shells and long reign of trilobites, like these Middle Cambrian *Ellipsocephalus,* mean they are common fossils found all over the world.

NEXT PAGE: The eyes of a scallop, seen here in this detail, are surprisingly complex, and can dilate and contract, so the shellfish is aware of what's going on in its environment.

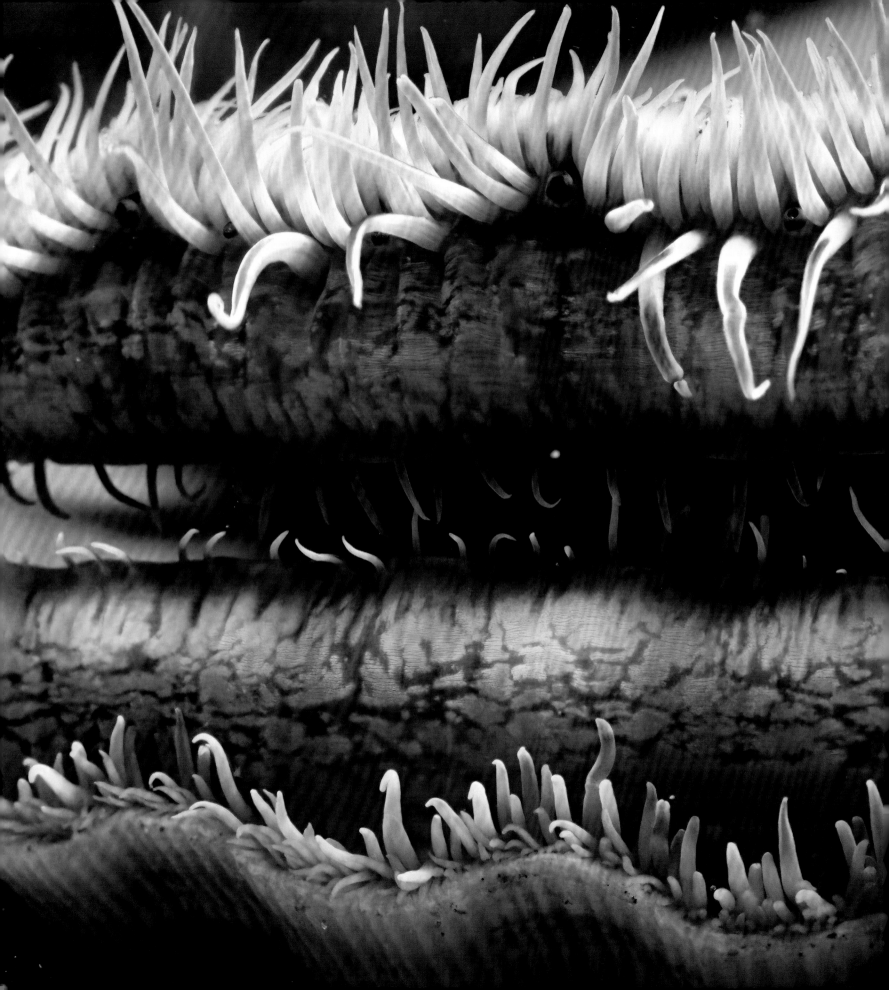

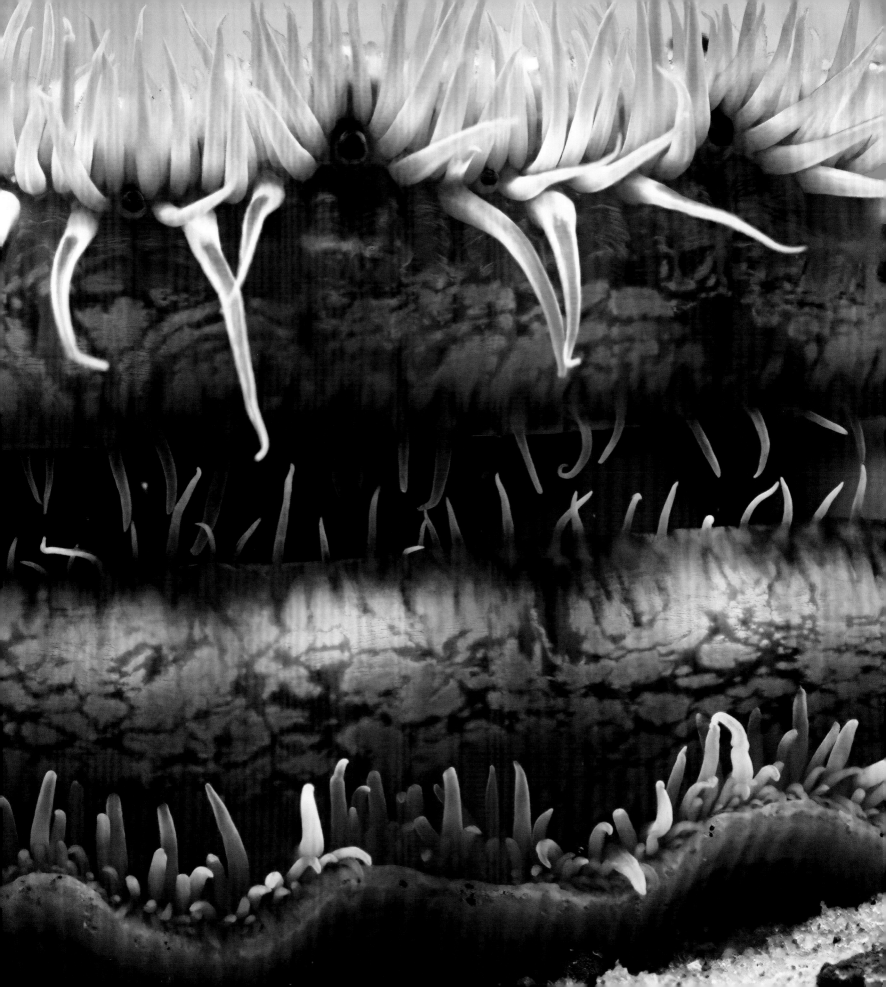

information about soft tissues, palaeontologists rely on the rare instances when conditions were just right for them to be preserved. This usually means that the animal has had to be buried before it could rot, and that the chemistry is just right to preserve the flesh, rather than just the hard parts. Very occasionally, this perfect storm of fortune produces some truly exceptional fossils – muscles, eyes, hair, feathers, internal organs and even colour can be preserved. In these circumstances our impression of an ecosystem can be completely transformed. If, for example, a modern sea floor was fossilised normally, we might only find crab and clam shells, with maybe some bones and teeth. However, if everything was preserved, we would see the worms, seaweed, anemones and all the other creatures that leave nothing behind when they die. These sites of exceptional fossilisation, where the rocks preserve incredible detail, are called Lagerstätte – German for 'storage place'.

The Chengjiang Lagerstätte continues to revolutionise our understanding of early animal evolution, with new discoveries announced every year. However, despite being known about for centuries, the importance of the Chengjiang was only recognised by scientists in 1984. Long before this, in 1909, palaeontologist Charles Doolittle Walcott had discovered exceptional Cambrian fossils in the mountains of British Columbia, Canada. These 508-million-year-old bands of grey, muddy rock are known as the Burgess Shale, and they have been captivating scientists for over a century. Among the life found there are arthropods, sponges, worms, molluscs and even a primitive chordate (our own group) called Pikaia. Among their ranks are some of the oddest creatures ever described, and many were completely new to science, with no modern relatives. Some even defied explanation entirely, such as the psychedelic-looking *Hallucigenia*, or the bizarre arthropod Opabinia, with its five stalked eyes. Much like the Chengjiang, these fossils preserved detail of soft tissues and provided a rare window into a complete Cambrian ecosystem. They also gave great insight into the soft anatomy of incomplete fossils found elsewhere.

One particularly common animal in Cambrian rocks around the world were trilobites – woodlouse-like arthropods with hard segmented exoskeletons. Normally, only their shells were fossilised, but the Burgess Shale preserved legs, gills and even a long pair of antennae. Decades of study and debate ensued as scientists attempted to classify and learn about these animals. Inevitably, ideas have changed as new evidence has emerged over the years. Take, for example, *Hallucigenia*, a Burgess Shale animal that was so unusual that it almost defied explanation. With nothing alive today to compare it to, scientists assumed that it walked on its long spines, however, further study revealed these spines were probably held aloft to defend itself while it

walked on soft tubular limbs, showing that previous reconstructions had the animals walking upside down.

The most fascinating story, though, is that of *Anomalocaris* – literally, 'abnormal shrimp'. The name itself comes from the first fossil that was found in 1892 – a dark, curved structure that looked like an odd kind of shrimp to the scientist describing it. Later, Charles Walcott found another piece of the animal, a pineapple-like ring of plates with spikes on the inner surface, which was thought at first to be a jellyfish. Later, another scientist found the middle part of the body but described it as a new species of sponge. It was only when a complete *Anomalocaris* fossil was discovered that the shrimp-, jellyfish- and sponge-shaped pieces were finally put together. An almost century-old mystery was solved, and in doing so it had revealed an incredible creature. *Anomalocaris* was a giant 40- to 70-centimetre-long arthropod that dwarfed all the

ABOVE: *Anomalocaris* was an early apex predator, with keen eyesight and spiked, grasping arms.

other animals. It had a streamlined body with paddle-like flaps along each side, which it used to swim. The head had two large, stalked eyes and at the very front it had two heavily spined, grasping arms — the 'odd shrimp' shape that gave *Anomalocaris* its name. The 'jellyfish' was, in fact, the mouth, complete with inwards-pointing spikes for pulverising food. Another remarkable feature of *Anomalocaris* was its powerful eyes, which had 16,000 lenses, and genetic studies of related animals show that it probably had colour vision as well. It is estimated that its eyesight was around 30 times better than that of a trilobite that lived at the same time. Without a doubt *Anomalocaris* was an apex predator, and trilobite armour would have been a key survival adaptation.

At around this time, 510 million years ago, trilobites were also beginning to experiment with enrollment, whereby they would curl into a ball shape so that only their hard outer shell was exposed, just as pill bugs do today. In many instances, trilobites were fossilised in this enrolled state, where they have panicked and then been buried alive, becoming entombed in sediment. Both of these strategies would have helped the trilobites survive in the dangerous waters of the Burgess Shale ecosystem. *Anomalocaris* and its relatives have now been identified all over the world, making it one of the world's first great predators. However, despite being

RIGHT: The Cambrian saw an arms race of defence and weaponry. Here *Anomalocaris* stalks an *Olenoides* trilobite in the ancient seas of what is now the Canadian Rockies, British Columbia.

lower on the food chain, trilobites were the true winners during this period. Thousands of species evolved worldwide, reaching peak diversity by the end of the Cambrian; testament to an astonishingly successful body plan that served them well for over 260 million years of Earth history.

Alongside trilobites, the molluscs were also diversifying rapidly, and familiar groups like gastropods (snails and their relatives), cephalopods (squid and octopus relatives), and bivalves (clams and mussels) all appeared at this time. The Cambrian was a revolution, transforming the quiet expanses of the seabed into bustling communities. Much like our own civilisations, organisms on the sea floor competed to live in the most productive areas, but the space available was strictly limited. Their solution, just like ours, was to build upwards. Organisms that required sunlight or food-laden currents were especially likely to grow taller or compete more fiercely for the best spots. Lifeforms not only carpeted the sea floor, they were crowding upwards into open water as well. The pressure was on to explore pastures new, and the land was beckoning.

BELOW: Among the first land pioneers were simple and low-growing organisms, such as lichens (left) and liverworts (right).

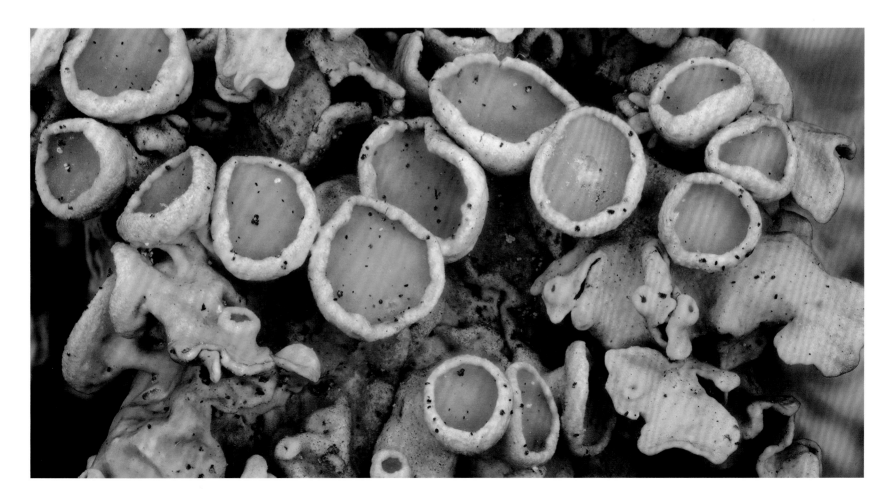

LAND PIONEERS

For billions of years the land had been a barren place, home to cyanobacteria and other microscopic life. By the middle Cambrian period, however, these crusts of simple life were beginning to become more interesting. Among the first lifeforms to creep onto land were lichens, which despite being plant-like are actually a strange symbiosis – algae living amongst filaments of fungi. As they spread, they helped to break down the rocks and produce some of the first soils. Today, they may cover close to 8 per cent of the Earth's land surface and are still amongst the first pioneers to grow on exposed bare rock. At around the same time, algae-like ancestors of land plants were slowly creeping up the rivers and shorelines. These early pioneers were simple and low-lying, probably similar to modern liverworts or green algae. Transitioning to a life on land would have been a huge challenge for these organisms, especially with the constant threat of drying out. For a long time they would have hugged the water's edge for moisture, and relied on water to help spread their reproductive spores. In time, plants would

evolve cuticles to reduce moisture loss and break their bonds with the ocean entirely, but for now, they would have to wait.

In hot pursuit of the first land plants and lichens were the arthropods, which faced similarly huge challenges in this hostile environment – not least breathing and supporting their body weight out of water. Even temporary forays onto the land would have given the first brave explorers an edge – escaping aquatic predators and feasting on the pioneering land plants and other animals. Whether these first terrestrial explorers were driven there by competition or drawn there by opportunity, we will never know. Indeed, the whole process of moving onto land is shrouded in mystery. After an animal dies on land, its body decomposes very quickly, and with no water to bury it, the chances of fossilisation are tiny. Without a good fossil record of the first land ecosystems, we have to once again rely on molecular clocks to fill in the gaps of our knowledge.

In 2022, a study using this method concluded that myriopods – the group that today contains millipedes and centipedes – were among the first to crawl onto the land. This is supported by the first physical fossil evidence, an air-breathing millipede called Pneumodesmus, which is found in much younger Silurian rocks. The authors of this study suggest the first land myriopods fed on the bacterial crusts and simple plants growing at the water's edge, and that predators, like arachnids, may have followed them there shortly after.

THE EDEN OF THE ORDOVICIAN

The Ordovician period, which followed the Cambrian 485 million years ago, was a time of global tectonic change. As continents shifted, volcanic eruptions pumped carbon dioxide into the atmosphere, elevating temperatures from pole to pole. Without ice locking up liquid water, higher sea levels flooded the land and created huge expanses of shallow-water habitat. Life continued to diversify at an incredible rate, with the number of marine families tripling by the end of the period. As well as the sheer amount of sea-floor space, the Ordovician world also offered a great variety of habitats, including the first coral reefs. Perhaps the most impressive denizens of the Ordovician reefs was a group of cephalopods called the nautiloids. Today's nautiloids are small, squid-like creatures that live on the slopes of coral reefs. Unlike most other living cephalopods, such as squid and octopus, they have coiled shells into which they can retract, rather like snails. The chambers of their shells are filled with gas, and by adjusting the levels, they can change their buoyancy by either floating upwards or sinking deeper. As well

as this vertical movement, they can propel themselves backwards with jets of water squirted out of a special tube called a siphuncle. Ordovician nautiloids had very similar anatomies and would have moved in a comparable way, but unlike their modern counterparts many grew to monstrous sizes. *Cameroceras*, for example, with its long, cone-shaped shell, could grow to over 6 metres across and would have been the largest animal on the reef by quite some way. Among its prey would have been the various invertebrates of the seabed, and perhaps even the odd fish – if they could catch them.

At the height of the Middle Ordovician diversification, fishes were small (around 25 centimetres long) and relatively simple, resembling armoured tadpoles. They lacked jaws, but did have

ABOVE: The giant-shelled cephalopod *Cameroceras* hunts for a trilobite amongst the rock and sponges of an Ordovician seaway 470 million years ago.

eyes, scales and lines of sensory pits along the body, which were used to detect pressure changes in the water. Jawless fishes today are quite rare, with only the hagfish and lampreys surviving, but they were incredibly successful in the Ordovician period. In order to swim, early jawless fishes, such as arandaspids, could beat their tails, but they lacked stabilising fins on their bodies for controlled movement. Later, some jawless fishes evolved fin-like extensions of their bodies to swim more effectively in the open water, but many others began to live on the sea floor. This was a dangerous place for a little fish; as well as giant nautiloids, like *Cameroceras*, a group of fearsome arthropods called eurypterids (or 'sea scorpions') were also evolving. As the name suggests, these were closely related to arachnids but, unlike their modern descendants, they typically grew to a metre in length, with huge, spiked claws. Perhaps unsurprisingly, many of the jawless fishes that followed were heavily armoured, choosing defence over speed.

The Ordovician had seen around 30 million years of rapid diversification, and the densely populated shallows were an Eden of productivity. Life had received its booster shot, and the seas had never been busier. Many trilobites became more elaborate and diverse, as did gastropods, corals, sea urchins, clams and many others. Where the Cambrian explosion had created the major branches of the tree of life, the Ordovician saw a growing of the twigs in between. All over the planet, every niche was occupied, from reefs to the deep oceans.

THE GREAT FREEZE

A dramatic change of fortune would put this expansion to an end. The high sea level that had flooded the continents and opened up so much habitat for shallow marine organisms was about to disappear. Around 445 million years ago, tropical sea temperatures plummeted by 10°C – the equivalent of cooling Barbados to Boston. As humans, we are all too aware of the discomfort that comes from even a small decrease in temperature, so for the species that had become fine-tuned over millions of years to live and thrive in warm waters, this was a disaster. Extinction was imminent.

What was the cause of the devastating drop in temperature? Volcanic gases, as we have seen, can drive global warming; however, volcanoes can also indirectly help remove greenhouse gases in the long term. When volcanic rocks such as lava are exposed to weathering, the silica-rich minerals they contain act like a chemical sponge for carbon dioxide. It is fascinating to think that the volcanic eruptions that warmed the planet with carbon dioxide and helped drive diversification may have later cooled it as the lava weathered away. Another intriguing explanation of the temperature drop is the spread of land plants at around this time. Primitive land plants surviving today, such as liverworts and mosses, are able to break down rock chemically as they grow, unlocking nutrients like iron, phosphorus and potassium. As these nutrients are washed out to sea, they can supercharge the growth of algae, which uses up vast amounts of carbon dioxide as it photosynthesises. A knock-on effect of these green blooms is that when algae die they begin to rot, and the water becomes stagnant as bacteria use up the dissolved oxygen. Great swathes of the ocean would have suffered, and at the worst time possible. Those that could adapt to the dropping temperatures and avoid the stagnant zones were faced by an even greater obstacle.

As the South Pole began to freeze, free liquid water was being locked away as ice, causing sea levels to drop by perhaps 100 metres. Those flooded continents, which had provided so much real estate for reef life, drained away, and every year the tides retreated further, leaving behind a dry graveyard of shells and skeletons. Amongst those victims, the large nautiloids were hit particularly hard. With the expanses of continental seas reduced to a fraction of their former glory, there simply wasn't anywhere to go.

This massive glaciation event may have lasted for hundreds of thousands of years, causing the extinction of 60–85 per cent of marine species. It was the first of the 'Big Five' mass-extinction events, and the second-largest in Earth's history. Then, slowly but surely, perhaps because of the eruption of more volcanic greenhouse gases, temperatures rose again. As the ice melted – gradually at first – ever more heat from the Sun could be absorbed by the oceans. Over time, currents could travel deeper into the fragmented ice sheets, accelerating the great melt until there was nothing left. Finally, the long road to recovery could begin, but ecosystems would take five million years to heal.

THE REEFS RECOVER

The next geological period, the Silurian, began 444 million years ago, and lasted for around 20 million years. It is named after an ancient Welsh tribe called the Silures, who occupied South Wales, the place where these geological divisions were first identified. In fact, three of these early geological periods reference Welsh rocks, because the Cambrian is named after the Latin word for 'Wales' (Cambria), and the Ordovician after another ancient tribe, the Ordovices, who lived north of the Silures. The warming, which had begun after the Ordovician extinction, continued into the Silurian period, when sea levels rose to reflood the continents. Sea levels would flutter up and down throughout this period, swinging by over 100 metres at times.

In general, this was a time of climatic stability and a welcome rest from the chaos. Shallow tropical seas dominated in the area of the Equator, allowing reef communities to recover. They were populated by new corals, sea lilies and little shelled creatures called brachiopods. Although patchy at first, rich ecosystems began to return, but the monster predators, like the nautiloids, had taken a backseat, which meant there was an opportunity to fill the role of apex predator – with fish poised to do just that. Jawless fishes were incredibly successful at this point, on both the sea floor and up in the water column. Many continued to experiment with fins and different tail shapes to improve their swimming. Among the bottom-dwellers were the charming ostracoderms, which had tank-like headshields and thick plates protecting the body.

Zipping around above them were the thelodonts and anaspids, strange jawless wonders unlike anything alive today. Although it wasn't obvious, there was an uprising in their midst that would change everything. Among the shoals of jawless fishes, a new group was becoming ever more common. Possibly as long ago as the end of the Ordovician, a branch of the jawless fishes had begun a remarkable journey. In order to breathe, fishes have structures in the head that

keep the gills splayed out like a fan, exposing them to as much water flow as possible. This improves gas exchange, so the fish can absorb oxygen dissolved in the water and get rid of waste carbon dioxide at the same time. These strong, rigid, support structures are called gill arches and can still be seen if you peer into the mouth of a modern fish. One group, probably the ostracoderms, began using the front gill arches to close their mouths. This would have been a weak motion at first, but eventually the arches evolved to become dedicated to biting with force, completing their transformation to become the first jaws. In the Silurian, the jawed fishes split into three groups; the first and most primitive being the placoderms, followed by the bony and cartilaginous fishes. During the Silurian, these two new lines of jawed fishes were only a small part of ecosystems, but their day would come. While this change was simmering away in the oceans, a more radical transformation was happening on land.

ABOVE: A striped mackerel shows its gills as it filter-feeds in the Red Sea, Egypt. Gill arches at the front of the mouth of early fishes had evolved into jaws by the Silurian period.

STRANGE FORESTS

By around 425 million years ago, plants were becoming more complex. A new group, known as the vascular plants, were growing taller, craning to the skies to maximise their access to sunlight. They faced gravity, physical damage from the elements, and the problem of how to transport fluid and nutrients around their taller bodies. Their secret weapon was a rigid polymer called lignin. By reinforcing their cell walls with this tough and watertight substance, plants could strengthen their tissues. This allowed vascular plants to support their own weight as they ventured upwards, and vastly improve fluid transport over these longer distances. The more primitive mosses and liverworts were left behind down below, but even in the shadows of the taller vascular plants, they remain incredible survivors to this day. It is important to note that the early vascular plants were not quite towering giants – *Cooksonia*, one of the first, was just a few centimetres tall. The largest organisms on land at this time were, in fact, a type of fungus called *Prototaxites*, which stood eight times the height of the tallest plants, with wide, tusk-like structures projected into the air, reaching over 8 metres in height and truly dominating the Silurian landscape.

Fungi may have originated in sediments below the sea floor around 2.4 billion years ago, before gradually creeping onto land in the form of lichens. The first forays of pure fungi on the land created a milestone for terrestrialisation, because of their profound effect on plants. They spread using extensive networks of tendril-like hyphae, which release enzymes capable of breaking down rock. In doing so, they can enrich soils and make nutrients such as phosphorus available to plants. This is a very direct process, and almost all vascular plants have an intimate symbiotic relationship with the fungi that are entangled in their roots. In return for nutrients, the plant provides the fungi with water and sugars it has produced via photosynthesis above ground. This is one of the oldest symbiotic relationships on Earth, and without it most plants would not survive.

Fungal networks are not only crucial to plant health, they are also truly immense. A handful of soil can contain well over 100 kilometres of fungal hyphae and make up half of its biomass. Fungi prevent the leaching of nutrients and moisture from soils and in addition are a significant sink for carbon. Despite their importance, though, we know remarkably little about fungi, and serious scientific interest in the group has only picked up very recently. The giant fungus *Prototaxites* emerged around 420 million years ago, standing proud amongst a miniature forest of early vascular plants. For the next 70 million years, it would witness a revolution on land, as waves of new invaders made it their home.

RIGHT: A *Favolaschia* fungus fruiting body emerges above ground to release its spores. Fungi form close partnerships with plants, and help recycle nutrients from decomposing wood.

INVADERS OF
THE LAND

The climate stability of the Silurian had allowed life in the seas to bounce back to its former diversity, after the first of Earth's 'Big Five' extinctions. The Devonian period that followed saw this diversification continue and complex food webs become established once more. At this point, around 419 million years ago, the supercontinent of Gondwana remained nestled over the South Pole, surrounded by smaller landmasses to the north. The planet was generally quite warm, which kept the land dry and desert-like, but the lack of ice at the poles also meant that sea levels were high. As a result, the jumble of island continents around the Equator were surrounded by warm shallow seas, offering huge amounts of habitat for reef organisms. Among these were the Palaeo-Tethys and Rheic Oceans, which would become the setting for one of the most aggressive arms races in Earth's history.

As the sunlit shallows nurtured the thriving reefs, they turbocharged the evolution of new species. The seas of the early Devonian would have looked quite different to those we know today. The surface temperatures hovered at around 25–30°C, which meant it was often too warm for corals to dominate. Instead, stromatolites, which we met in the first chapter, made up most of the reef structure. Huge mounds of these ancient microbial structures were peppered with clams, sea lilies, small-shelled creatures called brachiopods, and trilobites. Beyond the sea floor, the productive

PREVIOUS PAGE: Amidst a backdrop of fierce competition in the Devonian seas, many animals began to seek food and refuge on land.

LEFT: These *Gonioclymenia* ammonites were heavily defended, but they still needed to seek safety in numbers in the Devonian seaway.

Devonian oceans also saw many groups exploring the water column. Among them were the cephalopods, which had coiled their shells to become the first ammonoids. These ammonoids, although distantly related to nautiloids like the *Cameroceras*, were now much smaller and no longer top of the food chain. At the beginning of the Devonian, that title belonged to the eurypterids (sea scorpions), apex predators at the peak of their diversity. These heavily armoured arthropods were reaching over 2.5 metres long, armed with huge claws and excellent eyesight. Despite being such formidable animals, eurypterids saw a steep decline as the Devonian progressed. Any number of natural processes can affect how successful an animal group is over time, but the fall of the eurypterids occurred just as another group of predators was on the rise. Their competitors were our ancestors – the vertebrates.

AN AGE OF FISHES

From the earliest jawed fishes three groups had emerged: the placoderms, the cartilaginous and the bony fishes. For 30 million years they had been evolving steadily, in the shadows of predatory invertebrates such as sea scorpions and cephalopods. The Devonian, however, would see such an explosion in the fishes' diversity that it would become known by scientists as 'The Age of Fishes'. The most primitive of the three groups were the placoderms, heavily armoured fishes with bony plates and scales covering their bodies. While a few placoderms may have had true teeth, the majority used pairs of self-sharpening bony plates that formed beak-like structures. These were capable of shearing through the hard exoskeleton of eurypterids, and even the armour of other placoderms. Perhaps the most remarkable innovation of the placoderms, though, was internal.

For decades a Devonian reef called the Gogo Formation in Western Australia has been known for its fossil fishes. The preservation is so good that muscle tissue, nerves, a stomach, intestines, liver and even a heart have been found fossilised here. In 2008, a team of Australian fish researchers announced the discovery of a truly astonishing fossil (even by Gogo standards) – a 25-centimetre-long placoderm preserved with a baby inside and the umbilical cord and yolk sac still attached. Fittingly, this enchanting little creature was named *Materpiscis* (or 'mother fish'), and it remains the earliest-known vertebrate to show live birth. Similar finds in other species suggest that live birth might have been quite common in placoderms, showing an uncommon investment of time and energy in reproduction. Most modern fishes simply scatter eggs *en masse*, gambling on the fact that some will probably survive. In these cases, the male often releases sperm in close proximity to the female to fertilise the eggs. Placoderms, however, took this process a step further. There is evidence that males reproduced by inserting an appendage called a clasper into the female, just as sharks do today. By doing this, the male not only ensures that the sperm and egg meet, but also that he is the father of the resulting offspring.

Throughout the Devonian, placoderms diversified to fill a huge number of niches, tweaking their body plan to suit different ecological needs. There were bottom-living species that resembled stingrays and catfish (although they were not related), and even 7-metre-long whale-shark lookalikes which fed on plankton. In biology, this is called convergent evolution, where unrelated animals look the same as each other because they are living in similar ways. A great example of this is the similarity between sharks and dolphins. Despite their evolutionary paths splitting around 450 million years ago, they both have streamlined bodies, stabilising fins around the middle of the body, and powerful tails to produce thrust. This is because both groups have to overcome the same obstacles to survive, in both cases making swimming more efficient to catch prey. There are countless other

BELOW: *Coccosteus* was a placoderm, a group that was among the first jawed fish. Placoderms became very successful throughout the Devonian period.

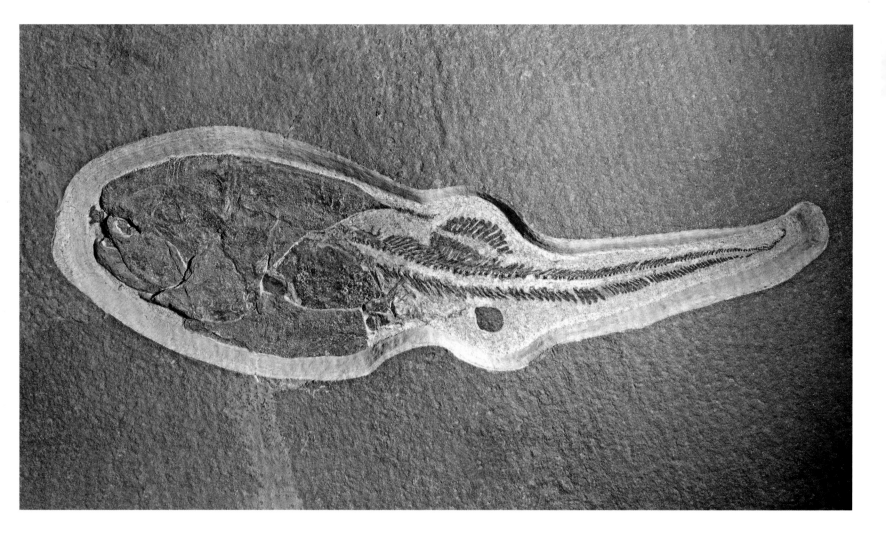

examples, which makes sense through the lens of shared problem-solving.

Placoderms, like the arthropods, had convergently evolved thick external armour for protection, but there was a price to pay. The defensive shield was a trade-off with speed and agility, and the ballast-like weight meant that many placoderms were restricted to life on or near the sea floor. Some, however – particularly the arthrodires – were fast and active hunters up in the water column. The most famous of these arthrodires is *Dunkleosteus*, a 9-metre-long superpredator of the Late Devonian. Not only was *Dunkleosteus* huge, but its jaws were nothing short of magnificent. The blade-like bony plates of the head were arranged in such a way that these fishes could bite down with a force of over 7,000 newtons, close to 500 times more than our own bite force and 20 times that of a saltwater crocodile, the modern world's largest and most powerful reptile. Even more incredible was the speed of attack, with one analysis suggesting *Dunkleosteus* could open its mouth in under a tenth of a second. This would have created huge suction, pulling prey into the mouth five times faster than a blink of an eye. Free-swimming prey, such as other armoured placoderms and thick-shelled ammonoids, did not stand a chance, but what of the other fishes in these terrifying waters?

Cartilaginous fishes – the ancestors of sharks and rays – continued to diversify

RIGHT: *Dunkleosteus* was the largest placoderm at, perhaps, 9 metres long, and even as a juvenile it could bite with astonishing force.

throughout the Devonian, evolving in many strange and interesting directions. Quite a few had very impressive defensive spines, which may have deterred predators trying to swallow them. However, at this time non-placoderm fishes were relying much more on speed to survive in the oceans and evolving a number of tricks to dart through the water efficiently. Despite being an important presence on the reefs, cartilaginous and bony fishes still very much played second fiddle to the placoderms and would do so for the rest of the Devonian – at least, in the oceans.

DAWN OF THE FORESTS

Since the very first plants crept onto land in the Ordovician period, they had slowly been adapting to its challenges. Chief amongst these was the constant danger of drying out, even in the most humid air. For a long time, the liverworts and mosses monopolised the moist lake and river margins, but it didn't take long for competition to arrive. By the Devonian, 390 million years ago, ferns, horsetails and even tree-like plants were taking root. The evolution of true roots and woody tissue allowed these plants to reach dizzying heights, with horsetails and other species reaching 30–40 metres tall. The vascular plants could not only grow higher into the air, but their roots were actively extracting nutrients from below the ground. Most remarkably, these plants were even transforming the landscape itself. As tangles of roots spread out to support the lofty growth, they also bound the soils together. This meant that soils were held in place on land, rather than being eroded and washed away by the rain. The plants also slowed down the progress of water over land, giving the soils that were building up time to soak up more moisture. Fast-moving torrents that had previously raced down from the mountains and out to sea became lazy, meandering rivers surrounded by lush forests.

Plants were engineering their own habitats and thriving but were still held back by one fundamental obstacle: in order to reproduce, they needed liquid water. Mosses, for example, can be male or female, and when the male's sperm reaches maturity it must swim through water to the female's eggs and fertilise them. After fertilisation, they would then produce embryos called spores, single cells that are vulnerable to drying out. Although primitive, this water-dependent lifestyle remains very successful for some plants even today. Mosses are found all over the globe and can resist desiccation quite effectively, even though their leaves are often only one cell thick. The reproductive capsules they produce can contain hundreds of thousands of spores, which are carried dust-like on the wind to grow elsewhere. It is a grand gamble, but the sheer number of spores means a few land in a moist place and survive. The next great leap forward would

come when vascular plants evolved a means to step away from the water for good … and the answer was seeds. Seeds were a safety capsule for plant embryos, complete with a food supply and protective casing. The earliest seeds in the fossil record even have wing-like projections, which could catch the wind and carry them for many kilometres. On impact, their protective seed coats would have shielded them from desiccation. Unlike the fragile spores, seeds could survive exposed, relying on their stores of starch and proteins, waiting in a state of suspended animation for the rains to come. Step by step, plants could now invade drier habitats and exploit the vast swathes of land beyond the wetlands.

This complex and expanding ecosystem was a rich habitat for arthropods and they continued to diversify. Among them were familiar groups like centipedes, millipedes, true scorpions and mites, as well as strange, spider-like arachnids known as trigonotarbids. They did not produce silk

ABOVE: Scorpions were among the first animals to leave the sea, over 430 million years ago.

like modern spiders, but instead they may have used their fangs and powerful venom to immobilise prey. The exoskeleton that had served so well as armour in the seas was now a perfect structural support for their body weight in air, and air was becoming increasingly oxygen rich as plants greened the world's land surface.

It is around this time that we have tantalising evidence for the first insects, but these discoveries are still controversial. Genetic studies have shown that insects had branched off to form their own group by the Silurian, so this is perfectly possible. These kinds of studies have also shed light on how insects fit into the arthropod family tree. Amazingly, it appears that insects are most closely related to crustaceans, specifically the fairy shrimps. The exact process of insects moving onto land is not clear, but by the end of the Devonian, the stage was set for them to thrive. Oxygen levels in the atmosphere had risen to 21 per cent, higher than at any point prior in Earth's history. The same could not be said for many freshwater habitats. The huge mass of plants growing and dying on land clogged and choked the waterways with rotting vegetation. As leaves and logs piled up, natural processes of decay used up oxygen dissolved in the water and were turning it stagnant. Any fishes living in these tea-stained waterworlds would need to get their oxygen from the air.

THE FISH OUT OF WATER

In the Devonian swamps, 400 million years ago, breathing air would have been a useful survival strategy, but, more than this, it was a vital step towards life out of water. The vertebrates' conquest of the land is one of the most enduring images of evolution: a brave fish venturing out of its ancestral home into a new world, and never looking back. Major hurdles lay in their way, which plants and arthropods had faced and overcome long before them. The most fundamental challenge was how would they breathe? A fish's gills are perfectly suited to exchange gases in water, with a large surface area well-supplied with blood vessels. The water is taken in through the mouth and constantly bellowed over these structures to ventilate them with fresh oxygen and carry away waste carbon dioxide. This is ideal in situations where dissolved oxygen is plentiful, such as shallow seas or fast-flowing rivers, but out of water, gills collapse under their own weight and become almost useless. Remarkably, many modern species that live in slow-moving and poorly oxygenated waters are capable of 'breathing'. Some, like the electric eel, have such poorly developed gills that they have to breathe air to survive and would, in a sense, drown if trapped underwater. Fishes breathe either by holding air in the mouth, swallowing it, or passing it into the swim bladder. This

RIGHT: Axolotls retain juvenile amphibian features, such as gills, into adulthood, making them fully aquatic.

sack-like organ is normally used to keep fishes upright, but its rich blood supply also makes it ideal for absorbing oxygen. What is more, the placement and plumbing of this bladder support the idea that it may have been the precursor to true lungs. You can see echoes of this transition during development in modern amphibians using gills in their tadpole form and lungs as adults. Some aquatic amphibians, like the axolotl, retain these gills throughout their lives but supplement their oxygen with an occasional gulp of air from the surface.

The exact timing of the shift from gills to lungs is quite difficult to pin down in the fossil record because these soft structures quickly rot away after the animal dies. Thankfully, there is much better fossil evidence for the evolution of fins to limbs, the key to efficient locomotion on land. Aquatic animals can maintain a neutral buoyancy because most of their body is the same density as the water surrounding them. The remaining denser body tissues (like skeletons) are subject to gravity and cause the animal to sink. Some animals prefer this arrangement and produce more heavy tissue to stay at the bottom, like many crustaceans, catfish and even hippos. However, for those that want to stay in the mid- or surface waters, there are a number of ways to keep from sinking. The most obvious is to actively swim against gravity like tuna do, but this requires a lot of energy. A more efficient way for less-active animals is to hold

LEFT: Male great blue spotted mudskippers fight for dominance on the shores of Kyushu Island, Japan. These fish survive on land by breathing through their skin and mouth lining.

a lightweight substance within the body. This could be a gas, such as that which fills fish swim bladders or the nautiloid's shell, or buoyant fats, as in the livers of sharks. In the air and out of the dense water, the organs of animals like sharks and squid are soon crushed by their own body weight. An ancient fish living in air would not only need to support its body out of water, but also travel while carrying its own weight. Any adaptation for living on land would first have to be useful in water. This begs the question, what is the use of an intermediate between fins and limbs? Modern fishes may hold the key. Over 200 living species of fish are amphibious and can use their fins to traverse land over short distances between water bodies. In most cases, this is an ungraceful flipflop of the entire body, but some, like the mudskipper, can remain upright while they heave forwards using their strong front fins. Underwater, a fish's pectoral fins (the equivalent of our arms) can be used to 'walk' along the bottom, and in some cases this is favoured over swimming. Stocky, well-muscled pectoral fins are perfect for grappling through dense vegetation and would have served ancient amphibious fishes well in the shallows.

The physical differences between water and air meant that fishes' eyesight had to change as well. Just before the transition to land, amphibious fishes' eyes tripled in size and moved from the sides of the head to the top. Bigger eyes don't make much of a difference to visual power underwater, but they can see much further through air. Having the eyes on top of the head like crocodiles meant that they could see prey above water while remaining submerged. Perfect for stealthy ambush predators looking for a meal on riverbanks.

Despite these obstacles and more, we know that slowly but surely fishes did make land their permanent home to become amphibians. As with the plants and arthropods before them, we will probably never know whether they were pushed there by competition or drawn there by opportunity. At the beginning of the Devonian, competition in the sea was certainly very fierce. This may have pushed bony fishes into the streams and wetlands, and by the end of the period even these freshwaters were becoming dangerous, which may have caused that final drive onto land. Throughout the amphibian's ascent, a group of fishes, known as the rhizodonts, ruled the swampy waterways. These animals were gigantic, with some reaching body lengths in excess of 7 metres, the largest freshwater fishes in Earth's history. Rhizodonts were, in fact, some of the earliest amphibian ancestors, but their branch of the family tree never left the water entirely. Like the earliest amphibians, rhizodonts possessed very well-developed, lobe-like fins to grapple their way through underwater vegetation. These would have been useful for making short journeys across the land, and perhaps even pursuing slower-moving creatures as they tried to escape the water. Rhizodonts were almost certainly ambush predators, with large fins at the back of their slender bodies for rapid acceleration like a dart. A strong bite combined with huge fangs at the front of their mouth would have been very effective for catching fish and even amphibians.

TOO MUCH OF A GOOD THING

By the end of the Devonian, the Earth's surface was green and lush. Plants had spread over the land from pole to pole, towered up into the skies, and photosynthesis was in overdrive. Huge volumes of carbon dioxide were being consumed and converted to carbohydrates, which was locked away in the plants' tissues. As carbon dioxide waned, so too did the temperatures, and Earth's climate became more and more unstable. What followed 374–358 million years ago was a series of devastating extinctions collectively known as the 'end-Devonian mass extinction'. What is puzzling about these extinctions is that organisms from shallow sea environments were particularly badly affected. Those on land appear to have been unfazed by these events, but why? Scientists are still debating the reasons for these pulses of death, but one particular theory has been gaining a lot of traction in recent times.

For millions of years the tropical Devonian forests and swamps had been producing rich soils. Roots had been liberating nutrients from the bedrock, including large amounts of phosphorus and iron. All the while the slow-moving waters had allowed thick piles of dead vegetation to build up, locked away and buried in stagnant pools. As the climate became increasingly unsettled, it is possible that these ecosystems became more unstable. More severe weather events and greater rainfall could have helped flush these swamplands, washing away the nutrient-rich soil into waterways and out to sea. The effect of all this fertiliser entering the sunlit seas was massive algal blooms, which would have turned the crystal-blue waters into a thick, pea-green soup of plankton. In the short term this would have blocked out the sunlight that was needed for photosynthesis by the organisms beneath, but as the algae died, a much more dangerous process began. The bacteria breaking down the decaying algae would have consumed huge amounts of oxygen, killing anything that could not breathe air.

Similar algal blooms happen today, where intensive agriculture on land results in fertilisers draining into rivers and finally into estuaries. The run-off causes algal blooms that can be seen from space and soon develop into huge dead zones. At the end of the Devonian, the seas would have been devastated by such influxes of nutrients. As oxygen levels plummeted in the shallows, anything that could not escape into the open seas or the furthest depths would have been vulnerable. Trilobites started evolving smaller eyes to live in the twilight of the deep as the shallow reefs ebbed away and corals suffered huge losses. Marine vertebrates were particularly badly hit, with an estimated 96 per cent of species wiped out by these events. Among the extinct groups were the placoderms, and that icon of the Devonian seas, *Dunkleosteus*. Bony and cartilaginous fishes hung on, perhaps able to swim away from the dead zones, or live at depth like the trilobites. Along with the placoderms, most of the jawless fishes, such as thelodonts, osteostracans and others, disappeared. Those fishes that did survive became much smaller, deprived of the food needed to sustain larger bodies. It was the second of the Big Five mass extinctions, with possibly over 80 per cent of species lost in total, victims in the sea of the land plants' success.

RECLAIMING THE WAVES

The Carboniferous period, which followed the extinction 358 million years ago, saw a power vacuum in the shallow seas. The monstrous armoured placoderms had left behind vacant ecological niches, and the surviving fishes soon began to fill them. A branch of the bony fishes, called ray-finned, began to diversify, alongside a suite of new cartilaginous fish, including the ancestors of sharks. The sharks experimented with a number of unusual body types throughout the Carboniferous, with some bordering on the bizarre. There were long and pointy eel-like forms, some with flatted disc-like bodies, and even species with huge pectoral fins that could be used to glide, like flying fish today. One group evolved spirals of teeth in the bottom jaw that resembled a circular saw, and would have been fierce hunters. Perhaps the most intriguing are *Stethacanthus*, which had a dorsal fin resembling an ironing board or anvil. When swimming, this would cause a large amount of drag,

ABOVE: The Devonian ended with mass extinction; stagnant, algae-choked seas saw the end of *Dunkleosteus* and this unfortunate ammonite.

so its function is still a mystery today. In the warm waters of the Carboniferous seas, the invertebrate recovery continued as well; sea urchins, sea lilies, trilobites, clams and gastropods had pulled through, along with shelled cephalopods like ammonoids and nautiloids. As productive a time as this was for the oceans, it paled in comparison to what was happening on land.

THE COAL SWAMPS

Plants like cycads, horsetails, club mosses, ferns and primitive conifers flourished during the Carboniferous. They formed sprawling forests that continued to consume carbon dioxide and drive down temperatures slowly but steadily. However, the air temperature during the Early Carboniferous was still a comfortable average of 20°C, and the great swamps dominated the landscape. Having evolved alongside the arthropods that made a meal of them, plants had responded by developing thick, lignin-rich bark. As leaves, branches and entire trunks fell into the brown, tannin-stained pools below, thick piles of carbon gradually accumulated. Over time these would become buried and solidify, compressed by the weight of the material on top to become coal. Today, most of the coal we burn for fuel comes from Carboniferous deposits, and the period's name literally means 'coal-bearing'.

As well as locking away huge amounts of carbon, the ongoing photosynthesis had a profound influence on oxygen. Levels in the atmosphere may have peaked at an incredible 35 per cent in the Carboniferous, compared to the 21 per cent we experience today. The combination of huge quantities of wood and oxygen made for a very volatile world. A single lightning strike could well have sparked an inferno, and wildfires were likely a very common sight at this time. Despite their flammability, these swamp forests were resilient and diverse, a paradise for invertebrates. The Carboniferous period saw the evolution of some quite remarkable arthropods, which were using the higher oxygen to their advantage. On the ground, millipedes grew to extraordinary sizes, such as the 2.6-metre-long *Arthropleura*. At 50 centimetres wide and built like a tank, this creature would have bulldozed its way through the forest with very few predators able to crack its thick armour. It was the largest millipede in history, living at a time with equally impressive record-breakers, among them the 70-centimetre-long scorpion *Pulmonoscorpius*, and the dragonfly relative *Meganeura* darting up into the air with a 60-centimetre wingspan.

This gigantism was all possible because of the higher oxygen levels on the planet. Land arthropods rely on oxygen reaching the muscle tissues efficiently through special breathing holes in the exoskeleton. As the body gets larger, this becomes more difficult, and to achieve the same

RIGHT: At 2.6 metres long and half a metre wide, *Arthropleura* was the largest land invertebrate in Earth's history.

NEXT PAGE: *Arthropleura* males may have used courtship behaviours to attract females, in ways similar to many millipedes today.

efficiency, oxygen levels entering the body must be higher in the first place. This isn't the only limit to size on land; gravity is a significant obstacle for those with heavy external skeletons. As body weight goes up, so too must the thickness of the exoskeleton to support it. Without the buoyancy effect of water, there is a mechanical limit to the maximum size that an arthropod can reach on land. Whether large or small, invertebrate diversity exploded during the Carboniferous, with the flying insects taking centre stage. They evolved all manner of adaptations for feeding on plants, especially the fleshier reproductive parts. For the first time, the swamps would have been filled with the sound of rustling and buzzing insects, with the occasional fish splash and amphibian call.

UP ON ALL FOURS

By the Late Devonian, around 365 million years ago, amphibians had evolved strong limbs with six to eight toes. Despite this, none was quite capable of walking fully on land – at least with any grace. Instead, they would have spent most of their time in water, behaving in much the same way as crocodiles do today. The story then becomes a little uncertain. Following the end-Devonian mass extinction, there is a period of about 15 million years with a very poor fossil record, which has confounded scientists for decades. One Early Carboniferous pioneer that bridges the gap is the 1-metre-long tetrapod *Pederpes*. The blueprint of everything to come was there, in the flesh. Unlike its predecessors, this creature, which in life resembled a salamander, was well-suited to walking and living on land. Not only did it have well-developed feet that pointed forwards, it also had five toes on its feet. Having five digits on our hands and feet is something we, perhaps, don't think about. It may be that extra digits are more trouble than they are worth, or perhaps that five was the minimum number required to function. Evolution has a tendency to remove unneeded body parts, and, in fact, many animals have lost digits since the Carboniferous. Horses are perhaps the most extreme example of this, today possessing only a single digit on each limb.

After *Pederpes*, much larger amphibians would come to dominate the waterways, including the 3-metre-long *Anthracosaurus*. This animal was as adept in water as it was on land, with a large newt-like tail and stocky arms and legs. It probably had thick skin to resist desiccation and would have actively explored the forests surrounding the water. There was good reason to do so, because at the same time as amphibians were gaining ground, the lands were drying. As trees continued to take in carbon dioxide from the atmosphere, temperatures continued to drop, and as this happened liquid water was slowly being locked away as ice. By the middle Carboniferous, ice caps were beginning to form at the South Pole for the first time in millions of years. The swamps' days were numbered and the amphibians were faced with a stark choice. They required water to reproduce, so that their gelatinous eggs could remain moist during development and the tadpoles could swim away after hatching. While *Anthracosaurus* and many of its kind were capable of venturing far away from the rivers and lakes, they were always bound to return, imprisoned by their lifecycle.

The strategy needed to break this bond with water was almost identical to the way plants had achieved it over 60 million years earlier. Amphibians began laying eggs that had a tougher membrane surrounding the vulnerable embryo, making them less prone to losing water. By around 340 million years, these so-called 'amniotic' eggs had breathable outer shells, which were leathery or hard so they could support themselves out of water. Within an amniotic egg, the embryo is fed by a yolk-sac as it develops in a fluid-filled sac called the amnion – essentially the embryo's own personal pond. The metamorphosis from tadpole to adult was no longer necessary; instead, it all happened within the safety of these little survival pods. Like the seeds of plants, amniotic eggs meant that the young could hatch away from water and begin to leapfrog across the world with successive generations.

This was the dawn of a new vertebrate dynasty – the reptiles. The first of these had evolved at the warm and wet tropical Equator near the shorelines. Armed with their tough scaly skin and amniotic eggs, they soon began dispersing to much harsher environments. With colder and drier conditions away from the swamps, there was also less food, and perhaps because of this the earliest reptiles could only grow to about the size of modern lizards. Almost as soon as this new dynasty had evolved, a split in their family tree produced two new bloodlines: the reptile line and

BELOW: The Carboniferous amphibian *Anthracosaurus* could walk well on land but, like most living amphibians, it still relied on water to reproduce.

the synapsids. The reptile line (sauropsids) was the stock for descendants such as turtles, lizards, crocodiles, dinosaurs and the like. Synapsids, on the other hand, may have looked superficially like their reptile cousins but would one day become the mammals.

This was a monumental fork in the road for reptile and mammal ancestors, which would get larger and diversify over the next 20 million years. Towards the end of the Carboniferous, synapsids were by far the most dominant land vertebrates, evolving stronger jaws than most other animals and large, fang-like, canine teeth. This was just as well because the temperature drop, which had been going on since vascular plants took hold, was approaching its peak. The ice sheet over the South Pole had grown so big that it spanned 50 degrees latitude. All of that water frozen away starved the land of moisture, and by around 305 million years ago, a tipping point was reached – the swamps dried up and entire forest ecosystems collapsed. The coal-forming rainforests, which had been a signature of the Carboniferous for so long, were now scattered, few and far between. The 'Age of Amphibians' was over.

PANGAEA

In the Permian period that followed, temperatures rose once again, but larger tectonic forces prevented the forest's recovery. Throughout the Carboniferous, the landmasses of Earth were colliding, and by the Permian they had merged into one giant supercontinent, known as Pangaea. It was so vast that it produced an unending land bridge from the North Pole to the South Pole, with enormous oceans on either side. A consequence of the formation of Pangaea was that rainclouds could no longer reach the deep interior. So even in areas where the temperature was ideal for recovery, it was too arid for most vegetation to take hold. As a result, plants that could cope with the dry climate dominated during this period, including ginkgoes, conifers and cycads.

The old guard of amphibians were now restricted to pockets of forest near the coasts, but the dry-adapted reptiles and synapsids were in their element and getting much larger. In the Early Permian, 295 million years ago, huge 3- to 4-metre-long synapsids like *Dimetrodon* and *Edaphosaurus* evolved. These creatures had large sails on their backs, presumably for thermoregulation, and by holding the sail in full sun they could warm up for the day's activities, just as lizards do today. Equally, if their body temperature was too high they could use the sails as giant radiators to cool down as well.

In time, the sail-backed synapsids gave way to a more advanced group, called the therapsids. These animals lived in higher latitudes than the sail-backs and, perhaps as a result, had progressively

LEFT: This female *Inostrancevia* was a gorgonopsid, robust and powerful predators of the Late Permian. Their descendants would eventually lead to the mammals.

become warm-blooded. Many were clearly excellent hunters as well, having increased the power of their jaws and begun to run in a much more agile way than anything before them. The more primitive sail-backs ran in much the same way as lizards do today – sprawling, with their limbs out to the sides of the body. The back legs produced all of the power, and the front legs were used to steer. To run, they would bend their body from left to right to swing their arms and legs into position. In contrast, the therapsids were beginning to bend their bodies forwards and backwards, in a much more mammal-like way, with their limbs swinging like pendulums beneath them.

One particularly impressive group of therapsids, which evolved in the Middle Permian, were the gorgonopsians, so-called because of their nightmarish appearance. These animals became apex predators during the Late Permian, growing over 3.5 metres in length, with large, sabre-like, canine teeth. They likely had a very good sense of smell and their large eyes would have served them well while hunting at night. Alongside the gorgons of the Late Permian was another group of more benign synapsids, called the dicynodonts. Most of these were small, about the size of a dog, with thick, leathery, dimpled skin. These herbivores rapidly diversified and spread, becoming incredibly abundant, perhaps due to their ability to eat almost anything. Their powerful beaks and claws would have made short work of tough vegetation like seeds and roots. There is also evidence that they were strong diggers, and dicynodonts have even been found fossilised inside their deep, spiralling burrows.

The large herbivore niche was occupied by members of the reptile line that thrived in this world, especially in the wet and tropical areas of Pangaea. Some, like the *Scutosaurus*, were large tank-like creatures, which could reach 3 metres tall and over a tonne in weight. *Scutosaurus* and its relatives were heavily armoured, with bony plates in the skin and thick, helmet-like skulls. This was almost certainly to defend themselves against the mammal-line gorgonopsians, which would have posed a constant threat.

SIBERIAN TRAPS

Around 252 million years ago, far beneath the feet of these strange Permian creatures, fingers of molten rock began probing the Earth's crust. Cracks and weaknesses gave way to pipes of magma surging relentlessly towards the surface, and baking everything it came into contact with. At 1,200°C, carbon-rich rocks underground combusted, releasing carbon dioxide that bellowed into the skies as the Earth bulged and fractured. When the lava did reach the surface of what is now Siberia, it marked the beginning of, perhaps, 100,000 years of huge eruptions. Rivers of molten basalt like those seen in Hawaii and Iceland today flooded the land, destroying everything in their path. The spikes of volcanic activity would have looked apocalyptic, as fire burst from huge fissures in the ground.

At first, these eruptions were of no real consequence to the Earth – a local problem in the far northern reaches of the planet. Nowadays, around 50–70 volcanoes erupt each year, and while they are exciting spectacles, most are relatively well-behaved. What made the Siberian Traps eruptions different was their sheer scale: five million square kilometres of land were flooded in lava, up to 4 kilometres deep in places. In the short term, the explosive violence produced clouds of ash and sulphur dioxide that encircled the globe. This would have prevented sunlight from reaching the Earth's surface and increased the reflectivity of the atmosphere, causing temperatures to plummet. The volcanic sulphur dioxide quickly dissolved in cloud water, creating sulphuric acid

rain, which was nearing the strength of vinegar. Plants below quickly browned and disintegrated in these toxic downpours, and before long their roots lost hold of the soil beneath them. Unbound as the plants died, this fertile earth eroded quickly and nutrients like phosphorus were washed away into rivers and out to sea. Flora rotted and ecosystems on land collapsed.

This was just the beginning. As the relentless eruptions continued, something much more powerful was happening – another volcanic gas, carbon dioxide, was building up in the atmosphere. In huge volumes, it trapped energy from the Sun against the Earth in a smothering heat. Inland, the heat was routinely spiking above 60°C, and wildfires became common. Habitats were destroyed as flames ripped through forests, killing herds of animals and releasing even more carbon into the skies. The worst was yet to come, because as the oceans warmed, a sleeping giant was stirring. On the sea floor, methane produced by bacteria had been building up for millions of years, held in place by a lattice of water molecules. These ice-like substances are called methane clathrates and are a normal part of the ocean system. However, now that temperatures were increasing rapidly, these clathrates began to melt, and methane began to bubble up to the surface.

Methane is another potent greenhouse gas, and its liberation was a disaster for the climate, rapidly accelerating global warming. Two massive pulses of heat pushed temperatures to soar on land and sea by around 15°C, with averages of 40°C across the globe. In the ocean, huge areas were becoming uninhabitable. Most marine animals cannot survive temperatures above 35°C for long, as their bodies slowly cook, and a great deal of metabolic energy is needed to prevent their proteins from denaturing. The temperature also caused a choking of the sea floor. Cold water is needed to draw down oxygen to the depths, and with temperatures rising this natural mixing was at a standstill. Higher temperatures resulted in less gas dissolving in the water, and even right at the surface fish would have been left gasping for oxygen.

At the same time, the influx of nutrients from the dying land were inundating the shallows, causing algal blooms. As they died, the decomposing algae soon turned the clear blue seas into a thick brown soup of decay. Rising carbon dioxide and sulphur dioxide levels also acidified the oceans, dissolving the shells and exoskeletons of marine creatures around the world. Life at the Equator became impossible, and any species that were able to fled to tiny refuges at the poles. On a global scale, these temperatures reduced currents around the planet to slow crawls, and as oxygen plummeted, great blue deserts were formed in the oceans. On land, great swathes of the Equator became dessicated graveyards, barren and lifeless. It was the most devastating extinction the Earth has ever seen; the pulses of destruction almost killed every animal, and an estimated 95 per cent of species were gone forever, their unique 3.5-billion-year stories at an end.

LEFT: Volcanic eruptions at the end of the Permian, known as the Siberian Traps, caused a runaway greenhouse effect and global devastation.

IN COLD BLOOD

The mass extinction of the end-Permian had obliterated all but 5 per cent of species on Earth, a grim milestone that marked the beginning of the Triassic period, 252 million years ago. The oceans had suffered particularly high losses during the extinction, with corals, snails, sea lilies and a number of other important invertebrate groups coming very close to oblivion. Sea scorpions and trilobites had not been so lucky – after well over 200 million years of evolution, their branches on the tree of life had been cut forever. With such massive devastation and so little life remaining, recovery for the planet would be a long and difficult process.

The first two million years of the Triassic were particularly cruel because, just as temperatures seemed to be returning to normal, another pulse of Siberian Traps volcanism began. Masses of greenhouse gases were released back into the atmosphere once more, and the pattern of destruction, seen at the end-Permian, was unleashed all over again. Just as life had been showing signs of recovery, it had been pushed back into the furnace. Those species that had survived the first extinction were well-equipped to handle the smaller waves that followed, but nevertheless, this delayed the healing process. Following extinction events, it normally takes around 100,000 years for ecosystems to heal. However, this event was unlike anything before or since, and it would take a punishing six million years for species numbers to get back to pre-extinction levels. With destruction came opportunity; the ecological niches that were previously filled with a diversity of Permian creatures were now vacant, and the race was on to occupy them.

The first animals to return to the barren graveyards of the sea floor were the clams, albeit as a fraction of the number of species that existed before. With resources so limited, they could not grow very big, but they were capable of filter-feeding in lower-oxygen waters, living life in the slow lane. Above them, the first to reclaim the open waters were the coil-shelled ammonites, which evolved quickly in response to the climate chaos. Clams and ammonoids had a monopoly over the shallow seas for the first two million years of the Triassic, before other animals began to return. Sharks and bony fishes that had sought refuge in the colder high latitudes were among the first vertebrates to come back, but much larger predators soon followed.

PREVIOUS PAGE:
A Fabian's lizard keeps an eye out for competitors. These salt flat lizards are native to Chile, and are exceptionally well-adapted to the arid conditions of the Atacama desert.

RIGHT: Insects suffered heavily across the Permian–Triassic boundary, losing one in three families to habitat destruction and climate chaos.

NEXT PAGE:
Juvenile *Lystrosaurus* jostle in the searing early Triassic heat of what is now South Africa. This species was hugely successful in the wake of the end-Permian mass extinction.

RETURN TO THE WATER

Despite the devastation, reptiles were quick to take advantage of the ocean's potential. Remarkably, within just a million years of the extinction, some reptiles were already beginning to evolve flippers and long, slender snouts to catch ammonites. These were among the first in an entirely new branch of reptiles, which would include some of the most iconic marine predators to

come. The earliest of these were the ancestors of ichthyosaurs, a group that resembled dolphins in their streamlined fish-like appearance. This is an example of convergent evolution, where a passing resemblance is the result of the environmental pressures of living in water, rather than any genetic relatedness. In a similar way, a number of reptiles evolved in the Middle Triassic that resembled turtles but were not directly related. In fact, at the time, the ancestors of true turtles were quite lizard-like in appearance, living on land but perhaps occasionally dipping their toes in lakes.

At around the same time as the first ichthyosaurs were evolving 251 million years ago, a number of long-necked marine reptiles were also taking to the water. Many had slender, streamlined bodies and may have used their limbs to swim in the shallows much like seals do today. Their needle-like teeth were well-suited to catching fish, and they may have used their long necks to flush out prey on the sea floor. One group of these early long-necked reptiles, the plesiosaurs, would enjoy particular success throughout the Triassic and beyond. These animals developed large, paddle-like flippers and strong muscles to power through the water. Without a living animal to observe, the way that these animals moved has puzzled scientists since the 1800s. Thanks to recent studies using water tanks and computer

RIGHT: *Lystrosaurus* gathering to feed on a patch of scrubby vegetation. They were equipped with strong beaks and tusk-like teeth for tackling tough stems and tubers.

models, we now know that plesiosaur swimming was completely unique in nature. The front pair of flippers would have been pulled downwards first, creating a wake in the water. The back pair of flippers would have been pulled down just at the right time to get a boost from this wake, helping thrust the animal forwards. This is a similar principle as geese flying in a V-shape, saving energy by flying in the wake of the bird in front. By the middle Triassic, paddle-flippered reptiles like these and the dolphin-like ichthyosaurs had well and truly conquered the waves and would rule the oceans for over 170 million years.

THE DISASTER SPECIES

On land the extinction had wiped around 70 per cent of vertebrate species off the face of the Earth, including the giant predatory gorgonopsians. Wildfires and acid rain had caused mass die-offs of flora, starving the herbivores that relied on it and the predators that hunted them. For a short time this was a veritable bonanza for organisms that thrived on decay. Chief amongst them were fungi, which for over 6,000 years would have coated the dead and dying trees in a white mass of mould. The loss of trees also meant that water could tear through landscapes unimpeded, and the slow-meandering streams were lost. Following the extinction, forests at high latitudes were not able to take hold for at least four million years, and the full recovery of complex land ecosystems would take even longer. At the Equator, where conditions were much harsher, this regrowth would take an incredible 15 million years.

The lack of vegetation in the earliest Triassic favoured smaller animals that did not require as much food to survive. The supercontinent of Pangaea was an unforgiving landscape, and as such the dominant flora was tough and dry-adapted. Today, plants that live in particularly dry environments tend to be woody with smaller leaves to prevent water loss, and they can store moisture in tubers and thickened stems. One animal particularly well suited to feeding on these plants was a synapsid that had made it through the end-Permian mass extinction, *Lystrosaurus*. This animal's turtle-like beak and tusks made short work of the desert flora, and its thick leathery skin and ability to dig served it well under the baking Triassic sun. Like clams in the ocean, *Lystrosaurus* had a good monopoly of these post-apocalyptic habitats – so much so that they are often referred to as a 'disaster species', organisms that thrive in the aftermath of extinction events. In some places, *Lystrosaurus* made up well over 90 per cent of the vertebrate population, and loose herds of these pig-sized creatures would have been a common sight at the time. Most other synapsids had not been so lucky, but amongst the survivors was the fox-sized *Thrinaxodon*. As well as the

large canines typical of synapsids, *Thrinaxodon* also had chewing teeth in its jaws, and better-developed muscles for biting and chewing. More primitive synapsids had an issue with prolonged chewing, because when food was in the mouth it blocked the airway, which meant they couldn't eat and breathe at the same time. *Thrinaxodon* was amongst the first to evolve a roof to the mouth cavity, which separated it from the nasal cavity above. This enabled it to breathe through its nose while chewing food, an ability we still enjoy today. This small and unassuming creature was one of a new breed of synapsid that would one day lead to mammals. Throughout the Triassic, these mammal ancestors, known as cynodonts, would acquire the traits needed to survive in an increasingly competitive world. The Permian had been a golden age for the synapsids, but the Triassic belonged to the reptiles.

AN AGE OF REPTILES

In any normal ecosystem, different roles are filled by species that are adapted to survive in those niches. Over time, competition within those ecological niches can cause even further specialisation, and animals become fine-tuned to do very particular things. For example, a herbivore without

ABOVE: The deserts of early Triassic Pangaea were truly vast, and much of the Earth would have been covered in brick-red sands stained with dry rusted iron, like the Namib of today.

competition would happily eat the plants that are the easiest to digest and most abundant in a habitat. As the population of that herbivore rises, competition for the food source increases and it begins to get rarer. Eventually, new species may evolve which specialise on foods that require very specific adaptations. A herbivore feeding on tubers, for example, may evolve strong digging claws and a robust skull capable of crushing the tough material, while another may evolve a longer neck to feed on the tender new growth of a tall tree.

This natural process of niche filling and specialisation is what drives diversity in a habitat. Evolution will always depend on the survival of the fittest, and if a species is filling a niche effectively it is very difficult for others to compete with it. Highly specialised organisms do what they do very well, and as long as the niche exists their genes will be selected for. Extinction events can quickly destroy those niches, and in doing so they expose the vulnerability that comes with being a highly specialised species. In our herbivore example above, imagine if competition had produced hundreds of different species, each specialised to feed on very particular plants. If those plants disappeared, the specialists would be forced to compete with each other for what is left. This is why species that are not specialised tend to fare better through extinction events; generalists are less reliant on a specific food source or habitat and are

RIGHT: An erythrosuchid rests in the midday sun in the Karoo of South Africa. Archosaur-line reptiles like these would be remain the top predators on Earth for the next 180 million years.

not burdened with any extreme adaptations that may set them back. Masters of none will always have their place in an ecosystem, and time and time again they have inherited the Earth. Extinction events through history, although destructive, have been powerful drivers of evolutionary change. They can serve to reset the diversity of animal groups, back to more conservative and generalist body plans. The more severe the extinction, the fewer survivors remain to seed the next wave of diversification. In this regard the end-Permian extinction was particularly potent. Throughout the Permian there had been plenty of synapsids filling ecological niches very effectively, but the extinction had wiped the slate clean.

Despite the early success of disaster species like *Lystrosaurus*, their dominance was relatively short-lived. A power vacuum following the mass extinction led to an explosion of diversity, as reptiles raced to fill vacant ecological niches. Among them were the ancestors of a group called the archosaurs, whose name means 'ruling reptiles'. By the Early Triassic, these carnivores had left behind the primitive sprawling posture of their predecessors and were now holding their legs upright under their bodies. This gave them a lethal combination of speed and control, which would have made them very effective hunters. Some of the earliest of these ruling reptiles were the erythrosuchids, crocodile-like reptiles that reached lengths of 5 metres. A fifth of the body length was their enormous head, which was deep, robust and full of teeth, resembling that of a carnivorous dinosaur. Erythrosuchids would have made easy meals of the *Lystrosaurus*, which – perhaps coincidentally – disappeared almost immediately upon the arrival of these monsters. After this, the surviving relatives of *Lystrosaurus* were fairly successful as larger herbivores but would remain firmly on the archosaurs' menu. This was well and truly the age of the reptiles.

Erythrosuchids were among the first of a huge dynasty of archosaurs that would follow, and by around 247 million years ago, two distinct bloodlines had appeared. One would eventually lead to the crocodiles and the other to dinosaurs and the flying pterosaurs. Throughout the rest of the Triassic period, the crocodile-line archosaurs would remain the largest carnivorous group. Many were similar to the erythrosuchids, but were much more agile, perhaps even running on two feet. Others were semi-aquatic, feeding in much the same way as crocodiles do today, although they were not directly related. Some even began to superficially resemble dinosaurs, walking upright on two feet with long necks and small heads. Among the more unusual of these crocodile-line archosaurs were undoubtedly the aetosaurs. These herbivorous animals could reach up to 6 metres in length, but they had very small, pointed heads. They defended themselves with bony rows of armour along their backs, and some were even armed with 30-centimetre-long spines.

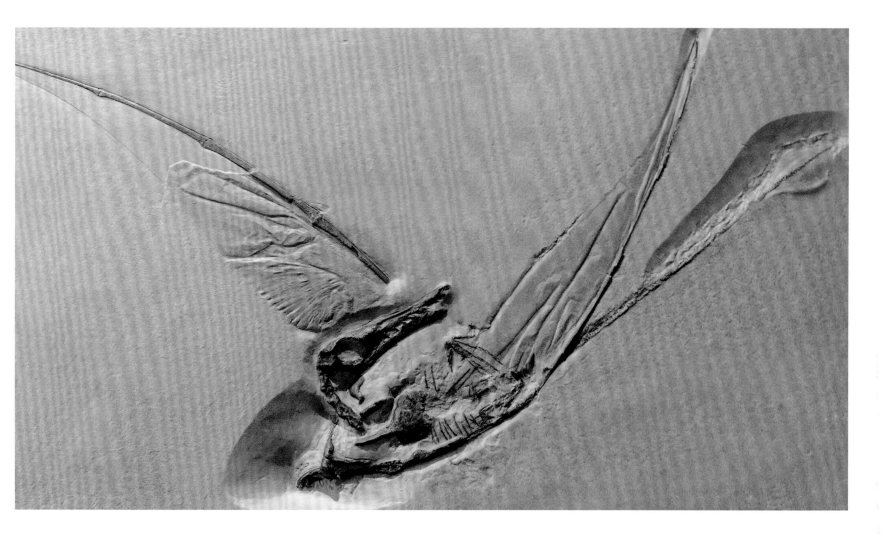

Arguably one of the most important groups of archosaurs in the Triassic were the rhynchosaurs, which originated in southwest Pangaea but soon spread worldwide. These herbivores were small but bulky, and superficially pig-like in appearance. They had a broad skull with powerful jaws and a cutting beak, and teeth that grew constantly to process tough vegetation. They became abundant in the Middle and Late Triassic, sometimes even making up 80 per cent of the animals in an ecosystem. Herds of these little archosaurs would have been a common sight worldwide, among a menagerie of other bizarre reptilian weirdos.

THE FIRST DINOSAURS

The branch of the archosaur family tree that led to pterosaurs and dinosaurs is undoubtedly the more famous, and more infamous than any other. As a result, this particular time in prehistory is subject to a huge amount of scientific debate. A number of key discoveries in the last decade have helped improve our understanding of this story enormously, but the details remain controversial.

ABOVE: This rhamphorhynchid was a primitive pterosaur. It had a long tail and large teeth. Fossils of fully winged pterosaurs first appear in the Late Triassic, but their early evolution is still a mystery.

The flying reptiles, the pterosaurs, probably branched off from the dinosaur line in the Early Triassic, although we are yet to find fossil evidence to narrow down exactly when. The first fossil pterosaurs are known from the Late Triassic, and at this point already had fully formed wings capable of flight. These wings were made of a leathery membrane that connected to the hind legs, giving the animals a superficially bat-like appearance. However, unlike bats, which use all of their fingers to support the wing membrane, pterosaur wings were supported by just one – the equivalent of our wedding-ring finger. The earliest pterosaurs had long tails and a mouth full of teeth, features that would be lost by later groups. The ability to fly would have made them highly effective hunters of small vertebrates and insects, and their strong claws show that they would have been excellent climbers as well. Many reptiles experimented with gliding during the Triassic, but pterosaurs were the first vertebrates in history to achieve true powered flight. They were so successful that they would monopolise the skies for over 70 million years, before the evolution of birds, the flying dinosaurs.

Like the pterosaurs, the story of early dinosaur evolution is still quite murky. Scientists agree that to qualify as a true dinosaur the animal must have at least three of their spinal discs (vertebrae) fused together into a structure called the sacrum. The sacrum helps to make a stronger connection between the backbone and hips, and this would have helped dinosaurs stabilise the hips while walking on two legs, something that all of the early dinosaurs share. By this definition, the earliest undisputed dinosaurs are found in the Middle Triassic around 233 million years ago. Before this point, there are tantalising glimpses of dinosaur-like archosaurs in the fossil record, but it isn't clear how they became true dinosaurs. Some of the earliest clues are footprints and skeletal remains, dating to perhaps over 245 million years ago. Among the best contenders for earliest dinosaur ancestor is *Nyasasaurus*, named after its discovery location, Lake Nyasa in Tanzania. The fossil was ignored for decades because it was quite scrappy, but, crucially, it does include a sacrum made up of three fused backbones. At 240 million years old, it is the oldest dinosaur yet described, although even earlier ones are almost certainly awaiting discovery. *Nyasasaurus* and its later descendants give us a good idea of what these early dinosaurs were like. They were probably small and active, running on two feet to catch prey, although they may have occasionally eaten vegetation as well. The lack of fossil dinosaurs at this early stage is probably a reflection of their rarity. Since their appearance, dinosaurs were relatively minor players in Middle Triassic ecosystems, and made up only 5–10 per cent of land animals. Their low abundance and diversity had slowly crept up as the Triassic continued, but they could not compete with the synapsids and crocodile-line archosaurs around them. Nothing suggested they were destined for greatness.

QUENCHING THE UNCROSSABLE DESERTS

The geography of Pangaea was a substantial obstacle for the early dinosaurs, which first evolved and diversified in the high latitude refuges of the south. The generally hot and arid conditions at the beginning of the Triassic meant that there were two huge belts of desert, lying to the north and south of a greener belt sandwiched at the Equator. So despite it being possible to walk across Pangaea from the south to north poles, it clearly wasn't an option for the dinosaurs to make that migration. It was only a quirk of geology that changed everything.

Around 234 million years ago, volcanoes in northern Pangaea, along what is now the west coast of Canada and Alaska, began to stir. Massive outpourings of lava – the so-called Wrangellia eruptions – followed, and with the molten rock came clouds of volcanic gases. First, sulphur dioxide and volcanic ash in the skies caused a snap of cold, then acid rains soon followed, which battered vegetation and destroyed the soils. A wave of rapid climate change then began as carbon

BELOW: The Carnian Pluvial Episode was a huge climate change event that drove moisture and life to the interior of the supercontinent Pangaea.

dioxide trapped heat in the atmosphere, raising global temperatures by 3°C. These shock changes in temperature and chemistry had devastating consequences for marine life. Dead zones spread, the sea floors became stagnant and delicate ecosystems collapsed as a third of the marine diversity was lost. Over a period of 1.5 million years, the planet experienced perhaps five of these climate fluctuations, and all the while soaring temperatures were changing weather systems as well. Warmer temperatures were driving evaporation from the oceans like never before, and water that went up, had to come down. Pangaea was so vast that moisture-laden air could seldom penetrate the interior, much like in central Asia and Africa today. However, mega monsoons were now becoming much more common, bringing unprecedented floods to the deserts. Year after year and millennia after millennia, the rain continued to rewrite the landscapes. In northern Pangaea, in a region that is now Scandinavia, these effects were particularly pronounced. Strong, moisture-laden trade winds were being forced over a 3,000-metre-high mountain ridge, and on the other side they were dropping an extraordinary amount of rain. So much, that an area of still lakes and land the size of Alaska was transformed into the largest river delta the Earth has ever seen. An order of magnitude larger than the Amazon basin.

As the moisture quenched the deserts, plants began to reclaim the land. Lush wet forests and vast river systems encroached, pushing back against the great desert walls. Conifers were among the first to take root, which adapted well to the harsh frontline. Huddled beneath the boughs of conifers, mosses, liverworts and other moisture-loving plants carpeted the once-parched landscape. This greening of the deserts finally allowed dinosaurs to disperse northwards, no longer trapped in the southern hemisphere. Pangaea was greener than it had been for tens of millions of years, but this period of intense rainfall was not to last. After just over a million years, the Earth's systems began to restore balance, and the climate slowly returned to its normal arid state. The lush, moisture-loving flora of the Pangaean interior began to die away as the rains became

LEFT: The Moenkopi Dinosaur Tracks, Arizona, USA. Animals leave behind many more footprints than bones and teeth in the fossil record, offering important information about how and where they lived and moved.

BELOW: The Late Triassic's *Herrerasaurus* of South America was one of the early dinosaurs that had a simple but effective body plan. It ran upright on two legs, powered by an efficient air-sac breathing system.

rarer. Once more, it was the deserts that grew, but with one significant change: the conifers were already adapted to dry habitats and weren't going anywhere. The wet spell had allowed them to take root across the planet and reclaim huge areas of desert, but the fabric of these forests was very different. Gone was the softer, leafier vegetation of the understory. Reptiles that could not reach or digest the fibrous plants were at a distinct disadvantage, as too were the ancient carnivores that fed on them. Animals like the rhynchosaurs became much rarer as the flora around them changed, and the rich green jungle gave way to sparse and towering conifers. These little herbivores would fade away in time, unable to access the bounty above their heads.

In contrast, the dinosaurs were thriving. By 232 million years ago, they had rocketed in abundance from around 20 per cent to 70 per cent of the planet's fauna. Among them were species like *Eoraptor* and its bigger cousin *Herrerasaurus*, which represented two major branches of the dinosaur family tree. *Eoraptor* was a small carnivore at around a metre long, but was nevertheless quick and agile. Despite its size, this bloodline would one day spawn the largest land animals that ever lived, the giant long-necked sauropods. *Herrerasaurus* was probably an early representative of the fearsome theropod dinosaurs, which would later include iconic predators like *Tyrannosaurus rex*. Fossils from around this time have also been attributed to a third branch of dinosaurs, the ornithischians, which eventually incorporated genera such as *Triceratops* and *Stegosaurus*. However, there is still very lively debate about whether they are true ornithischians or other similar-looking archosaurs. The jury is still out on this, because many of these early dinosaurs and non-dinosaur archosaurs look very similar.

A WINNING FORMULA

Following the wet interval 232 million years ago, dinosaurs became an important part of ecosystems worldwide. This begs the question, what made them so special? The reasons for the dinosaurs' ultimate success are still debated but several factors may have given them the competitive edge. Firstly, as we saw earlier, their hips had modified in a way such that the legs were held beneath the body as they walked and ran. While this certainly improved their agility, it also had advantages for respiration. Sprawling animals, like lizards, bend their bodies to swing the legs around while walking, and in doing so they squeeze the lung of one side of the body. Then, when they bend to bring the opposite legs around, the other lung gets compressed, and so on. Each step compresses a lung, making breathing near impossible when they are running. While all animals (including humans) can rely on anaerobic respiration for a short while, they cannot run for extended periods.

This is why lizards run in the way they do – in short bursts followed by frequent resting stops. The upright stance of dinosaurs, and indeed the mammal-line synapsids, solved this problem. By keeping the legs beneath the body and moving them in such a way that does not compress the chest, these animals could run and breathe at the same time.

Another adaptation that may have served the dinosaurs well was warm-bloodedness. For a long time after their discovery, the typical depiction of dinosaurs was of cold-blooded lumbering reptiles. However, by comparing the microscopic structure of their bones with modern animals, scientists now know that they were warm-blooded. By elevating their body above the ambient temperature, dinosaurs could remain active for longer. In contrast, cold-blooded creatures, like lizards, rely on the Sun's warmth to raise their body temperatures, and are quite sluggish without this extra energy. What is extraordinary is that it is now believed that many of the Triassic archosaurs and synapsids may have also been, at least, partially warm-blooded. As well as physical evidence, scientists can use a technique called 'bracketing' to deduce information about extinct animals. The idea is that if two species share an adaptation, their common ancestor probably had the same one. For example, if chimpanzees and humans both have hair, the long-extinct ape that

ABOVE: Lizards like this Yellow Spotted Monitor walk with a sprawling gait, limiting their ability to move and breathe at the same time.

gave rise to humans and chimps probably had hair as well. So, in the case of warm-bloodedness, we know that mammals have it, as do dinosaurs too, which means their ancient ancestors probably did as well. It is a blunt but powerful tool, especially when trying to work out how extinct animals may have behaved.

Warm-bloodedness may have been the trigger for another adaptation of the dinosaur-line reptiles: feathers. Higher body temperatures are of no use if the heat is lost through the skin, so animals like birds and mammals had to have some form of insulation. Hair, feathers and reptile scales are made of the protein keratin, and all three form in similar ways, growing from follicles in the skin. Unlike scales, which are flat and blocky, hair and feathers are long and fibrous, which helps to trap air against the skin. By doing this they can retain heat radiating from the body, and keep the animal at peak temperature and performance, day or night. Beautifully preserved dinosaurs from China have well-developed feathers all over the body, but no direct evidence has been found of this in Triassic dinosaurs. However, we do know that pterosaurs had simple filament-like feathers as well, which means we can once again apply the technique of bracketing. If we know that later dinosaurs had feathers and the closely related pterosaurs also had feathers, then it is very likely that the first dinosaurs had them as well. In the earliest stages, dinosaur feathers were probably very simple, little more than downy fluff to keep them warm.

Another trait that may have given dinosaurs the edge after the rains was their quite remarkable breathing system. Mammals breathe by inhaling air into the lungs and exhaling it out in the same direction. This mixes stale air with new every time they refill their lungs. To make matters worse, there is a period of time when breathing out that no gas exchange is taking place. So, in effect, the lungs are only being used for half of the breathing cycle. Dinosaurs, on the other hand, had a much more efficient system, which is shared by crocodiles and birds today. When dinosaurs breathed in, they would inflate giant air sacs within the body, and these would flush air through the lungs in one direction before it was breathed out again. These air sacs acted like bellows, constantly ventilating the lungs so that there was no pause to the gas exchange, and because this air was moving in one direction through the lungs, stale air was always being replaced by new. These air sacs were connected to a network of cavities in the skeleton, which would have helped to lighten their entire body. It was an incredibly efficient system, and one that would have fuelled a very active lifestyle.

GIANTS OF THE LATE TRIASSIC

It is still unclear which, if any, of these adaptations helped dinosaurs blossom over other vertebrate groups. It may have been a mix of opportunity following the climate disaster, or a combination of adaptations that gave them a competitive advantage. Whatever the reason, dinosaurs diversified rapidly over the next 30 million years, and by the Late Triassic they dominated the lands. Herbivores were particularly successful, and many were getting very large. Some of the most impressive were the sauropod ancestors, such as *Plateosaurus*, which evolved around 214 million years ago in the forests of what is now Europe. These enormous animals reached 10 metres in length and possibly 4 tonnes in weight. They walked on two feet, just like the earliest dinosaurs, counterbalancing the long flexible neck with a muscular tail. *Plateosaurus* had serrated teeth shaped like a leaf, much as iguanas do today. These were perfectly suited to both crush and slice plant material as the animal

LEFT: Feathers, like those of marabou storks, evolved originally to insulate body heat, much like the hair of mammals.

NEXT PAGE: Baby *Plateosaurus* were probably left to fend for themselves on the forest floor. They had to grow quickly to avoid larger predators.

fed. Some herbivores today, including deer, cows, hippos and even tortoises, have been observed supplementing their diet with meat, demonstrating that feeding strategies can be quite flexible.

This has been suggested for *Plateosaurus*, which may have occasionally eaten smaller animals if given the opportunity. *Plateosaurus* probably used its robust arms to grapple with the tree branches on which they fed, while their long necks would have given them a distinct advantage, reaching particularly high leaves, over smaller herbivores at the time. Compared to meat, plants do not offer much energy for the effort involved, which begs the question how could herbivores like *Plateosaurus* grow so large? Vertebrates cannot digest plant matter on their own. Instead, they rely on bacteria that produce enzymes capable of breaking down tough compounds like cellulose. The small head and large body of *Plateosaurus* suggest it used fermentation in its gut to process bulky plant matter, just as many herbivores do today. Crucially, when animals get bigger their digestive time goes up as well. If food is retained in the body for longer, the bacteria are given more time to break down the cellulose, and so more nutrients can be absorbed by the animal. It may be that very early sauropod ancestors, such as *Plateosaurus*, relied more on chewing to unlock the contents of plant cells, but in time the larger sauropods relied more

RIGHT: *Plateosaurus* were the first giant dinosaurs, reaching 10 metres in length. They fed mainly on plants, but may have supplemented their diet by scavenging meat as well.

and more on their fermentation chambers to do the hard work. Of course, the other great advantage of herbivores being large is that they are less likely to be targeted by hungry predators, and *Plateosaurus* had its share of enemies in the Late Triassic forests. Theropod dinosaurs had also become larger in the Late Triassic, like the 5-metre-long *Liliensternus*, which with its sharp backward-pointing teeth would have made short work of a younger *Plateosaurus*.

One strategy that all dinosaurs used to survive their vulnerable youth was to grow very large very quickly. Then, as they reached a certain critical size that was less attractive to predators, this growth rate would slow down. The huge sizes that dinosaurs were reaching was also a reflection of their conquest of the habitats. A rule of biology, which still applies today, is that species tend to get larger if the resources can support them. This is because larger body size gives animals a competitive advantage. They can, for example, move greater distances to feed, dominate more territory and resist more predators. The opposite is true when resources are limited. Generations become smaller because it is harder for them to find enough food to fuel larger bodies. By the end of the Triassic, dinosaurs were getting very large, but they were dwarfed by the giants of the ocean.

Throughout the Triassic a marine revolution was underway. The marine reptiles had diversified explosively in the first 20 million years of the Triassic, quickly exploiting all sorts of ecological niches. Some began chasing down fast-moving prey like fishes in the open water, evolving long slender snouts with evenly spaced teeth like a crocodile. Others were using more robust, crushing teeth to tackle hard-shelled ammonites, as well as clams on the sea floor. By the end of the Triassic, ichthyosaurs were reaching the peak of their diversity and some were getting very large. The North American species *Shonisaurus*, for example, reached lengths of over 20 metres. As well as a slender body and snout, these animals also had very large eyes, supported by a ring of bones in the socket. This suggests they were hunting in the darkness at depth, possibly feeding on squid-like cephalopods called belemnites that lived at the time.

END-TRIASSIC EXTINCTION

The Triassic period had begun 252 million years ago amidst the greatest extinction event in Earth's history. Volcanism in Siberia had driven global climate change and almost destroyed life on our planet. For 50 million years, the survivors had spawned a huge diversity of forms on land and in the oceans, building up stable ecosystems from nothing. Meanwhile, the Pac-Man-shaped supercontinent Pangaea remained Earth's only landmass, and lazily drifted north throughout the

LEFT: Like the beginning of the Triassic, the end of the period was marked with violent bouts of volcanism, climate warming and mass extinction.

Triassic. All seemed calm, but that was not to last. This period would end just as it had begun, with volcanic violence and climate chaos. Beneath Pangaea, a constant flow of circulating magma continued to push and pull on the Earth's crust. By the end of the Triassic, 202 million years ago, the huge tectonic forces wrestling below were becoming much stronger. A tug of war pulled Pangaea in two directions, north and south. The supercontinent began to groan with the strain, causing earthquakes and devastation above ground. The earthquakes became more frequent and more intense as the crust buckled and neared breaking point. Cracks appeared and grew larger, and so magma from below began seeping upwards, forcing the rock apart like a crowbar. These 'intrusions' of magma came in waves lasting over 50,000 years, baking carbon-rich rocks and releasing huge amounts of carbon dioxide as they did. After around a million years of these pulses below the surface, lava began erupting on the surface of what is now Nova Scotia and Morocco. In the heart of the Pangaea, this was the beginning of 600,000 years of volcanic activity. During this time, 3 million cubic kilometres of lava would erupt, covering an area the size of Europe. It was the most extensive outpouring of lava in Earth's history and would come to be known as the Central Atlantic Magmatic Province, or CAMP for short.

With this volcanism, came a now all too familiar chain of events. Just like the end-Permian extinction, the short term effects of volcanism were global cooling. Volcanic ejecta, rich in sulphur, clouded the atmosphere and blocked out the Sun. At higher latitudes, it was especially cold, and a mild glaciation took hold. As the volcanic sulphur dioxide dissolved in the clouds, it formed sulphuric acid, which rained down onto the wilting vegetation below. The plant roots, which bound the soils together, soon rotted away, destabilising the earth and allowing the phosphorus-rich muds to wash out to the sea. This influx of what was essentially fertiliser caused huge algal blooms in the oceans. Oxygen in the water plummeted as bacteria feasted on the decaying algae, choking swathes of shallow-water habitat. All the while, carbon dioxide was steadily rising to 7–8 times our modern levels, causing global warming of 4°C. As with the end-Permian, this may have triggered the melting and release of methane clathrates on the sea floor, making the greenhouse conditions even worse.

Carbon dioxide was also having a significant chemical effect on marine organisms. As carbon dioxide dissolves in water, it produces a weak acid, and higher concentrations of carbon dioxide in the atmosphere would have caused the oceans to acidify. Many marine animals make their skeleton out of calcium carbonate, which breaks down as it reacts chemically with this acid. The same sort of reaction that happens if you mix bicarbonate of soda with vinegar, but at a much slower rate. Animals that couldn't build their carbonate skeletons faster than they dissolved away in these acidic waters were doomed. Corals, clams and ammonites all suffered and, in total, a third of marine diversity was lost forever. Reef communities collapsed, and with them disappeared their top predators, the largest of the ichthyosaurs.

On land, the warmer global temperatures were causing the interior of Pangaea to become even hotter and drier. However, at the same time this heat was supercharging evaporation of water from the oceans, causing more powerful monsoon systems. As a result, some land areas began experiencing intense seasonal rains and a wetter climate. What followed was a massive turnover of local floras and restructuring of habitats. Plants, adapted to either wet or dry conditions, jostled for dominance in the rapidly changing ecosystems. This turmoil in climate, and rapid increases in global temperature 201 million years ago, was a disaster for land vertebrates. In just 10,000 years, they lost over 40 per cent of their diversity. Among the victims were many of the large amphibians that had hung on throughout the Triassic. The crocodile-line of archosaurs

BELOW: Many of the archosaur reptiles died out during the Triassic–Jurassic extinction, but the ancestors of crocodiles managed to survive the climate chaos.

was particularly badly hit, and the armoured aetosaurs and large carnivorous forms all disappeared. The era of strange and wonderful reptiles was at an end, and so the only survivors of this line of archosaurs were the crocodiles themselves.

The herbivorous synapsids, distant relatives of *Lystrosaurus*, were still lingering at the end of the Triassic, but were a shadow of their former glory. The end-Triassic extinction was the final straw for this ancient group, unable to cope with climate-driven changes in the flora around them. The only synapsids that would make it through were the cynodonts, ancestors of all mammals to come. By the end of the Triassic, these creatures were small and shrew-like in appearance, and probably nocturnal. They were also warm-blooded, which may have helped them survive the extinction event. As well as global warming, the volcanic bursts would have also created short cold snaps as aerosols and volcanic dust entered the atmosphere, blotting out the sun. There is even evidence of short glaciations at very high latitudes during this time, and so warm bloodedness and insulating hair would have certainly helped. This may explain the survival of other iconic animals as well. The dinosaurs and flying pterosaurs were relatively unphased by the end-Triassic extinction, perhaps because they had similar insulation, in the form of feathers.

The Triassic period had ended as it had begun, in a chaos of volcanism and global warming. Pangaea was finally beginning to split apart, with Laurasia to the north made up of what would become North America and Eurasia, and Gondwana to the south. Gondwana dominated the southern hemisphere and was truly immense, made up of what is now Antarctica, South America, Africa, Australia and others. Between Laurasia and Gondwana, the young Atlantic Ocean was getting wider, flooding what were the deserts of central Pangaea. The violent outpourings of lava which had accompanied the Atlantic's birth had now subsided, the climate was stabilising, and the waves of extinction were over. Life could, once again, begin to recover as it entered the most famous geological period, the Jurassic.

LEFT: The Central Atlantic Magmatic Province was a rift of volcanism that caused Pangaea to break apart. It allowed water to flood between the newly formed continents: Laurasia to the north and Gondwana to the south.

THE RISE OF THE DINOSAURS

The Jurassic period began 201 million years ago, in the midst of huge tectonic upheaval. The supercontinent of Pangaea continued to be torn apart, with volcanoes rumbling along the seams. The worst of the eruptions, which had caused rapid climate change and extinction at the end of the Triassic, were over, but Jurassic temperatures remained warm by modern standards. There were no permanent ice sheets at the poles, but a combination of tectonic activity and higher temperatures pushed up sea levels throughout the Jurassic. With no ice caps and therefore nothing to melt, why did sea levels rise during this warming period? The answer lies in the properties of water itself.

Like many materials, water expands when it gets warmer; the molecules are given more energy to vibrate and they exert a force outwards. On a hot day, railway tracks and roads buckle as they expand, and water acts in much the same way, occupying a larger volume at higher temperatures. Around a third of sea-level rise in recent times has been caused by this thermal expansion of water, and it is especially pronounced in the deep sea at higher pressures. In addition to this, the sea floor itself was changing, as masses of molten rock forced its way to the surface, causing the crust to bulge and buckle. Both of these effects would have been enough to raise global sea levels, even without meltwater from the poles.

A benefit of these higher global temperatures was that reefs crept up into higher latitudes than they are today and increased their range. At first, these reefs were made of sponges, stromatolites and worms that made hard tubes, but over time the more familiar corals took over. As Pangaea fragmented, water began to flood into the gaps between the newly forming continents. Weather systems could finally penetrate the innermost desert regions, and humid forests slowly took over the landscape. Even at the poles there were forests, with cycads, conifers, ginkgoes and tree ferns all thriving. Swathes of new Early Jurassic forest and freshwater became home to the first frogs, sturgeon fish and butterfly ancestors.

PREVIOUS PAGE: The giant herbivores *Triceratops* and *Alamosaurus* feed in a sunlit North American valley.

RIGHT: Following the Triassic–Jurassic extinction crocodile-like archosaurs found particular success in the oceans.

THE EXODUS TO THE SEAS

Alongside newly evolved groups, these productive habitats were quickly seized upon by the survivors of the extinction as well, including turtles and lizards. Although most of the large predatory crocodile-line archosaurs were gone, a single group did survive, one that would eventually lead to true crocodiles. These crocodile ancestors occupied some of the small to midsize predator niches left over from the Triassic, but in time these would be outcompeted by the dinosaurs. One group of crocodile ancestors, the thalattosuchians, would even become

marine, evolving all the traits of an aquatic predator throughout the Jurassic. They evolved tail fins and paddles to swim, and powerful jaws capable of attacking much-larger prey. At 3 to 7 metres long, thalattosuchians were undoubtedly apex predators at the time. However, they were entering the seas at a time of revolution, joining two other formidable groups of marine reptile.

Ichthyosaurs had also made it through the end-Triassic extinction, but along the way they suffered a loss of diversity. The surviving ichthyosaurs were, at most, half the size of their gigantic 20-metre-long ancestors, but they were nevertheless fearsome hunters. Almost all had a compact, dolphin-like body that was perfect for bursts of speed underwater – ideal for

hunting cephalopods and fishes. Many Jurassic ichthyosaurs evolved particularly large eyes, with some reaching 25 centimetres in diameter – bigger than a basketball and the largest of any animal in history. These eyes were so big that they even had a ring of special bones to support them, suggesting they were adapted for hunting in dim light, perhaps at great depths.

Like ichthyosaurs, the four-paddled plesiosaurs had evolved an aquatic lifestyle during the Triassic and they gave birth underwater. By the Jurassic, they had split into two groups, then they diversified rapidly to become top predators of the shallows and open water. Broadly, there were the long-necked plesiosaurs, which had small heads filled with needle-like teeth, and the pliosaurs, which had large, robust heads and relatively short necks. These reptiles grew to enormous sizes, occupying the large-predator niches left behind by the extinct Triassic ichthyosaurs. Exactly how these animals fed is still a subject of debate – especially the function

ABOVE: A Late Jurassic *Pliosaurus* stalks the shallow waters of Europe. Their streamlined shape and strong flippers made them particularly fast and manoeuvrable.

of the long-neck plesiosaurs. It is possible that the neck was used to root out fishes and other animals from small spaces, or perhaps as a way to plough the sea floor while swimming above it. Others have suggested that by having the head at the end of a long neck, the plesiosaurs could surprise schools of fish without giving away their true body size. The pliosaurs had relatively short necks, suggesting they were pursuit predators, using their strong paddle-fins to achieve sustained high speeds. Throughout the Jurassic, a number of pliosaurs attained sizes of over 10 metres, with their heads alone being in excess of 2 metres. Among them was *Pliosaurus funkei*, evocatively nicknamed 'Predator X' before its formal scientific description, and with good reason. *Pliosaurus* had well over 100 teeth in its crocodile-like skull, with some being up to 30 centimetres long. Remarkably, the size of the jaw muscles suggest a bite force twice that of *Tyrannosaurus rex*, and possibly the largest of any animal that has ever lived. This animal would have made short work of almost anything it laid its eyes on.

REVOLUTION IN THE REEFS

The bustling new reefs below the marine reptiles were also undergoing a revolution. As sea levels rose and water drowned the land, habitats from all sides of Pangaea began joining up. As this happened, species that had been separated for millions of years were reunited and thrown into conflict.

During the Early Jurassic, the first crab-like forms evolved from more lobster-like ancestors, perhaps in response to predation from marine reptiles. Crab-shaped organisms have evolved several times throughout Earth's history, where the meaty and vulnerable tail section of the animal becomes shorter and gets tucked under the body. The Jurassic also saw the evolution of hermit crabs, which could effectively double their defences by using the shells of other animals, such as ammonites. Ammonites themselves diversified rapidly throughout the Jurassic, and became very common all over the world. They evolved so quickly and were so widespread that they became a very useful tool for geologists. Layers of rock can look very similar, so it is difficult to compare them from place to place, however, by looking at the unique shell patterns of ammonite species, it was possible to work out the age of the rocks they came from. As people compared more outcrops of rocks and ammonites, maps were drawn that could be used by builders, miners, farmers — in fact, anyone interested in what was below the ground. A number of fossil groups can be used in this way, helping scientists reconstruct the story of life from rocks all over the planet.

ABOVE: *Eparietites* was an ammonite that thrived in Jurassic and Cretaceous seas all over the world, a time when many of their shells became particularly well suited to life in deeper waters. This specimen is 20 centimetres across.

Before fossilisation was fully understood, ammonites were a thing of folklore and mystery around the world, and their name comes from the coiled horns of the Greek god Ammon. In seventh-century England their shells were compared to serpents and thus became known as snakestones. Early Christians associated snakes with the Devil, and it was believed that saints of England had rid the sacred land of snakes by turning them into stone. Local collectors even carved heads onto the ammonite fossils to make them look more like snakes. Another group of cephalopods – the belemnites – were also becoming very common during the Jurassic, and had developed their own folklore. In life, the animal resembled a squid, and like their modern relatives they were an important food source for larger predators. The only hard part of the animal was a bullet-shaped, crystalline structure, which is often all that remains as a fossil. Again, their name comes from the Greek, this time the word *belemnon*, which means 'javelin' or 'dart'. They were believed to be remnants of thunderstorms that were flung down to Earth. In the Medieval period of Europe, the fossils were even placed in the roofs of houses to prevent lightning strikes. Both belemnites and ammonites were thought to have healing properties, including treating cramp in cows and encouraging them to produce more milk.

The greatest revolution in Jurassic waters, perhaps, was the diversification of fishes, including a group of bony fishes called ray-fins, which had been steadily taking over freshwater and marine habitats. Among these was a family called the pachycormids, which was an incredibly diverse and successful group of ray-finned fishes. Long before true tuna and swordfishes evolved, the pachycormids had species which looked near identical; another case of convergent evolution. These fishes would have been incredibly fast-moving and aggressive, rivalling the speed and agility of marlin and other game fishes today. However, even the most accomplished fisherman would have trouble catching the largest pachycormid, with conservative body length estimates of over 15 metres, *Leedsichthys* was the largest fish of the Jurassic seas and one of the largest in Earth's history. Today's record-breaking fish is the whale shark, at almost 19 metres, which can attain such huge sizes because of its diet of plankton and small fish. Like the whale shark, *Leedsichthys* was a filter-feeder, using its enlarged gills to extract plankton from the water. Filter-feeding pachycormids like *Leedsichthys* would occupy this ecological niche for 100 million years before their extinction at the end of the Cretaceous. It would be quite some time before large filter-feeding sharks and true whales filled the void and took their place in the modern oceans.

DINOSAURS RISING

By the end of the Triassic, dinosaurs were already relatively diverse, with a variety of body types and lifestyles. Perhaps because of their diversity they had survived the turmoil of the end-Triassic extinction, although any number of their unique adaptations could have made the difference. Dinosaurs had feathers, could run quickly and had efficient breathing systems, and the herbivores were well-suited to the tough vegetation of an arid greenhouse. With the majority of Triassic crocodile-line archosaurs gone, dinosaurs quickly monopolised habitats all over the world. Unlike any other herbivores, these animals were actively browsing the highest vegetation and had unique access to this food source. By the Jurassic, this had allowed sauropod ancestors to get much bigger, with some reaching around 10 metres in length. The necks

in particular began to get longer relative to the body, enabling them to reach higher up than their competitors. The bounty of food up in the treetops was enough to drive an evolutionary race skywards, but this process was not without its compromises. As with earlier sauropod ancestors, all this vegetation was probably being digested in huge fermentation vats in the gut. Over time, the animals became heavier and the necks got longer, making the whole body less stable and walking on two legs impossible. As a result, they began walking on all four legs, which became strong and pillar-like, much like those of elephants today. These were the first true sauropods, an iconic and instantly recognisable group of dinosaurs.

Sauropods are often depicted on screen as gentle and lumbering giants, but this is probably not the case in reality. Many living herbivores are very aggressive and defensive and do not take much encouragement to attack anything they perceive to be a threat. For example, in Africa many more people are killed by elephants and rhinos every year than by lions. There is no reason to think sauropods were any more benign, especially with large theropods stalking below them. The idea that they are slow-moving is probably borne from the fact that they are such huge animals, requiring a lot of power to move their heavy skeletons. However, studies of the neck bones have

LEFT: A young *Diplodocus* watches a dung beetle on the Late Jurassic floodplains of North America. This tiny dinosaur would one day grow to around 25 metres long.

NEXT PAGE: Dinosaurs of the Late Cretaceous Hell Creek Formation of North America, with giant pterosaurs flying above.

shown that they were remarkably lightweight, with around 60 per cent filled with air spaces. These air spaces not only made the neck easier to move quickly, but it also helped sauropods to breathe. Sacs connected to these air spaces were all part of the one-way breathing system common to all dinosaurs, and they helped bring in as much oxygen as possible. The long and lightweight neck was not just for reaching higher vegetation, it also allowed some of these animals to feed from the ground. By swinging the head around in an arc the animal could save energy by standing in one place, essentially using its long neck as a conveyor belt to the fermentation chambers. Having a relatively narrow neck and small head also meant that sauropods could reach for many metres into dense forests, where they couldn't squeeze their wide bodies.

Relatives of the Triassic theropod dinosaur *Liliensternus* had also made it through the extinction, and quickly began to diversify in the Early Jurassic. The best-known of these is *Dilophosaurus*, made famous by its venom-spitting antics in the film *Jurassic Park*. While this hunting strategy is unlikely, there is no doubt that *Dilophosaurus* was a fearsome predator. At around 7 metres long, it was one of the largest land carnivores of the time, and its slender jaws were filled with curved and serrated teeth. It would have been an agile and fast-running animal, capable of hunting a wide range of prey. *Dilophosaurus* did not have a neck frill, as in the film, but it did have an equally impressive crest on its head. This may have been brightly coloured, perhaps used for individuals to recognise one another, or for sexual displays, as in birds today. The Early Jurassic soon saw the emergence of a new lineage of larger theropods, called the tetanurans. This was a pivotal moment in theropod evolution because the tetanurans would diversify for over 120 million years, ultimately accounting for the vast majority of predatory dinosaurs and modern birds. Like the sauropods, the early tetanurans became very large, with many of the first species reaching well over 6 metres. One of the first of these, and the first to be discovered, was *Megalosaurus*.

LEFT: *Araucarias*, or 'monkey puzzle trees', of the Conguillío National Park, Chile. These gymnosperms would have been an important food source for long-necked sauropod dinosaurs.

BELOW: The lower jaw of *Megalosaurus*, a large carnivorous theropod dinosaur from the Middle Jurassic of England.

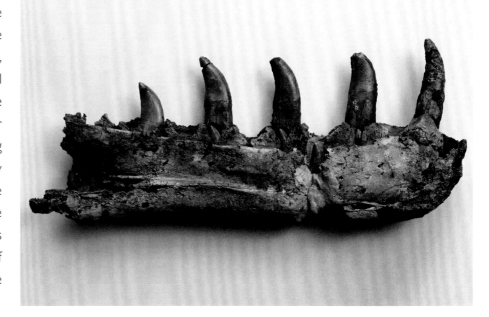

TERRIBLE LIZARDS AND SEA MONSTERS

Dinosaur fossils have been known for millennia, but they have often been interpreted as dragons and other mythical creatures. *Megalosaurus* was probably the first dinosaur to be scientifically described and named, although the first fossil discovered was only a fragment of leg bone. This piece was discovered in England in the mid-1600s, and was originally believed to belong to an elephant, then later a giant human. Unfortunately, the specimen bore a resemblance to male genitalia, and so illustrations of it were captioned with the name *Scrotum humanum*. This name could very well have stuck, but thankfully it was decided that this, humanity's first dinosaur discovery, would be called *Megalosaurus bucklandii* instead. The second half of the name is in honour of William Buckland, who described later fossil discoveries of *Megalosaurus* in more detail. As other dinosaurs were found, they caught the attention of anatomist Richard Owen. It was Owen who coined the name 'dinosaurs', meaning 'terrible lizards', after he realised that these massive reptiles were a new and distinct group of animals.

The 1800s was a time of great progress in natural sciences, and many strange creatures were being collected from the rocks of Europe. In the south of England, along what is now known as the Jurassic Coast, fossil collector Mary Anning had become a self-taught expert on marine reptiles. She made a living selling fossils to tourists and wealthy collectors, and is thought to be the inspiration for the tongue-twister 'she sells sea shells on the sea shore'. Many scientists, including William Buckland and Richard Owen, would flock to the area to consult with Anning about the fossils, and would collect them alongside her. However, at the time women were excluded formally from academic circles and Mary Anning was often not credited for her work in scientific papers, much to her frustration. Throughout her life, Anning discovered ichthyosaurs, plesiosaurs, a pterosaur and a number of other extinct animals that attracted huge interest. The idea that monstrous sea creatures once patrolled a quaint seaside resort captured the public's imagination and helped spark the frenzy of interest in prehistoric life, which continues to this day.

In the late 1800s this interest turned from Europe to North America, where many more dinosaurs were being unearthed. Two scientists of this time became particularly famous for their research: Othniel Charles Marsh and Edward Drinker Cope. At first their relationship was friendly and they even named new species after each other. However, it didn't take long for them to fall out, and then one of the most famous scientific feuds in history began. They competed with each other for the new fossils being unearthed, and criticised each other very publicly. The two raced to publish their finds, using whatever means they had at their disposal; Marsh was

better funded and would have tonnes of material delivered to him from an army of workers scouring the American West, while Cope bought a scientific journal so that he could rush his publications into print. The feud would only die with the men themselves, finally putting an end to what became known as the 'Bone Wars'. The reckless excavation and rushed descriptions of Cope and Marsh's fossils are still causing problems for scientists today, but it was by all accounts a golden age for palaeontology. Some of the most famous dinosaurs were discovered by these men, including *Triceratops*, *Stegosaurus*, *Allosaurus*, *Diplodocus* and many more.

ABOVE: The Crystal Palace sculptures in London are some of the oldest reconstructions of dinosaurs, and inspired huge public interest when they were unveiled in 1854.

ORNITHISCHIANS AND THE DINOSAUR FAMILY TREE

Sauropods and theropods were diversifying quickly in the Early Jurassic, filling the large herbivore and carnivore niches respectively. As sauropods reached higher into the leafy canopies for food, a new kind of dinosaur, called 'ornithischians', had evolved to feed on vegetation and even small animals below. These early ornithischians had serrated, chisel-like teeth and beaks, and some even had tusk-like canines that may have been useful for prising apart tough plant material. Unlike sauropods and theropods, ornithischians also had a pelvic bone, the pubis, that pointed backwards towards the tail rather than inwards towards the belly. This pelvic arrangement meant that there was more room for a larger gut, further helping with digestion.

Ornithischians would become some of the most important and numerous herbivores on land, but despite this their origins are still frustratingly unclear. Some palaeontologists believe ornithischians may have evolved before the Jurassic in the Triassic, at around the same time as the sauropod ancestors and theropods. The exact relationship of ornithischians to other dinosaurs is even more controversial. The traditional textbook idea is that the ancestor of all dinosaurs split into two in the Triassic – the ornithischians on one side, with the sauropods and theropods grouped together on the other. Very recently, however, some scientists have suggested that sauropods were the ones that split away from theropods and ornithischians. This debate will undoubtedly continue until better evidence is found from those earliest stages of dinosaur evolution. Whatever they are related to, ornithischians only really started to diversify and become successful in the Early Jurassic. One of the first groups to branch away from the earliest ornithischians were the armoured thyreophorans. Thyreophorans had bony plates along their backs and their skin was probably thick and leathery, covered in spikes and lumps. As they evolved, this armour became even more robust, and like many dinosaurs of the time they soon began to get much larger.

BELOW: Like primitive mammals of the Jurassic and Cretaceous, modern ocelots are well adapted to hunting after dark, with exceptional night vision and hearing.

LIFE IN THE SHADOWS

Throughout the Triassic, mammal-line synapsids had been doing a good job of staying out of the way of large reptilian predators. An effective way of doing this was to limit activities to the nighttime, becoming small, shrew-like and warm-blooded. Throughout the Late Triassic and Early Jurassic, their behaviour and anatomy became fine-tuned to this nocturnal existence. This included developing keen senses of touch and smell, and their eyesight became much stronger in the dark. When light enters the eye, it hits a layer of light-sensitive cells called the retina, which acts much like the film of a camera. Mammals improved their night vision by losing cells in the retina that perceive colour, in favour of those that collect more light. So while many mammals are colour blind today, they are still able to detect much more light, which enables them to see in very dim light. This is enhanced by another adaptation; many nocturnal mammals also have a reflective layer of tissue behind the retina of their eyes called the tapetum lucidum, which bounces light back through the retina to maximise detection. You may have seen this layer in the glowing eyes of modern mammals, such as cats at night.

Hearing was particularly advanced in early mammals, and the story of its evolution is quite extraordinary. Early mammal ancestors had a lower jaw made up of four bones, and a further two bones connecting that jaw to the skull. With this arrangement they were limited to hearing low-frequency sounds, like rumbles and deep growls. They would have heard the world entirely through their jaws. As the true mammals evolved, some of the bones around the jaw joint began to get smaller, broadening the range of sound frequencies they could hear. These bones then began to migrate further into the skull, and for a time were used for both hearing and chewing. Gradually, these bones became separated from the jaw entirely, forming the three-part middle ear arrangement familiar to us today. These little middle ear bones are fully enclosed and transmit vibrations from the eardrum to the inner ear. This remarkable change allowed mammal ancestors to hear progressively higher-pitched sounds, which would have been very useful when hunting insects at night. For a long time, mammals would remain in the shadows, burrowing underground or seeking refuge in the trees. The Early Jurassic saw a good diversity of these small mammals, and while they may not have been making any headlines, they survived perfectly well in their hidden niches.

CHOKING THE EARTH

The Early Jurassic had seen a steady recovery of the number of species, with diversity beginning to approach the peaks of the Triassic. However, this came to a sudden halt 183 million years ago, when life faced yet another volcanic hurdle. During the late Triassic, the vast continent of Pangaea had split entirely into two giant continents – Laurasia to the north and Gondwana to the south – and now Gondwana itself was beginning to break apart. Africa and Antarctica began to split, opening volcanic rifts in the crust that flooded the surface with 2.5 million cubic kilometres of lava. With it came over 20,000 gigatonnes of greenhouse gases, either bubbling out of the lava itself or baked away from carbon-rich rocks underground. The familiar domino effect of climate havoc began once more, and as carbon dioxide levels rose, so did global temperature – this time by around 5°C. This slowed down the mixing currents of the oceans, preventing oxygen from ventilating the depths and choking any life that lived on the sea floor. As carbon dioxide dissolved into the water, the sea became more acidic. The calcium carbonate skeletons of marine species became thinner and weaker as they lost the fight against the changing chemistry. Water acidification was particularly devastating for corals and many sponges, and in some places over half of marine species were lost.

On land the rising temperatures were causing aridity and the hardy drought-resistant vegetation was faring much better than wet-adapted plants like ferns. The rich and diverse forests were replaced by tough drought survivors like cycads and conifers. It was a relatively brief moment of floral

RIGHT: Lava pours into the Indian Ocean as vegetation burns near Saint-Philippe on Réunion Island – a common scene throughout prehistory.

change, but enough to fundamentally alter the composition of the communities of animals that relied on them. The greatest losers on land were the primitive smaller sauropods, which could no longer compete with the larger and more advanced species. These newer sauropod giants weighed over 5 tonnes, with long necks, robust skulls and jaws, and large teeth capable of processing the tougher vegetation. Their size allowed them to ferment larger quantities of food, extracting as much nutrition as possible as they bulk-fed on the tallest trees. It is unclear quite how other dinosaur groups responded to the changing character of the forests, but there were undoubtedly knock-on effects for other herbivores and their predators.

DRIFTING APART

The Middle Jurassic saw a number of important branching events in the family tree of theropod dinosaurs. First, the more primitive *Megalosaurus*-like theropods formed a new allosaur line, and in turn this split to produce the coelurosaurs. The earliest coelurosaurs of the Middle Jurassic included distant ancestors of *Tyrannosaurus*, although at this stage they were relatively small, at 3 to 5 metres in length. For the rest of the Jurassic, the large-predator niches were occupied by the three-fingered allosaur-type

RIGHT: Under a full moon, a herd of *Diplodocus* feed on ferns and high-growing pine needles. These large sauropods were particularly successful in North America during the Late Jurassic.

dinosaurs, including *Allosaurus* itself, which could reach around 10 metres. Ornithischians were also becoming more diverse in the Middle Jurassic; the heavily armoured thyreophorans had given rise to familiar groups like stegosaurs, ankylosaurs and an offshoot known as the ornithopods, which would one day include the duck-billed hadrosaurs. Later in the Jurassic, another split occurred from the ornithopods, which would eventually lead to dinosaurs like *Triceratops* and its relatives. Unfortunately, scientists are quite fond of their jargon, but putting all of that aside, it is clear that the Middle to Late Jurassic was a crucial stage of dinosaur evolution, with all sorts of new groups emerging.

The reason for this accelerated diversification is not entirely clear, but it is possible that the Earth itself was playing a significant role. As Pangaea continued to split apart throughout the Jurassic, the drifting landmasses took populations of dinosaurs with them. New, smaller continents with no land bridges to connect them meant that the dinosaurs of each province became genetically isolated. Over time, this may have led to new species evolving in a number of separate pockets around the world, causing diversity to soar. In addition, rising sea levels caused by global warming could have produced even more of these genetic islands.

PTEROSAURS

RIGHT: Just as in the Jurassic, the modern forests of China's Yunnan Province are a biodiversity hotspot, where complex habitats drive the evolution of new species.

NEXT PAGE: Pterosaurs, such as the Late Jurassic *Pterodactylus*, ruled the skies long before birds evolved. These flying reptiles had simple feathers, which they may have used for thermal insulation.

Reptiles ruled the Jurassic land and sea, and above them the flying pterosaurs were masters of the sky. Their ability to migrate or, perhaps, their feather-like coats, had helped them survive the end-Triassic extinction, and despite all the climate instability that followed they became much more diverse throughout the Jurassic. By the Middle Jurassic, 170 million years ago, pterosaurs were beginning to get much larger. A species described in 2022 from Scotland, called *Dearc*, had a wingspan of 2.5 metres or more and was not yet fully grown. Like other pterosaurs of the time, *Dearc* had a long tail, curved teeth in its jaws and large claws which would have helped it to climb. On the ground, *Dearc,* and others like it, would have moved quite awkwardly, because the broad wing membrane was attaching to the legs. Soon, however, a new breed of pterosaur would evolve with longer narrower wings, and with the skin membrane joining at the sides of the body. This left the legs free to move more easily, enabling the new pterosaurs to walk effectively on all fours, and even take off directly from the ground. For a while, the old and new pterosaurs lived side by side, but eventually the more free-wing design won out. The ability to fly was a powerful tool that had given the pterosaurs complete monopoly of the skies – but this wasn't to last.

CHINESE FORESTS

Around 160 million years ago, a huge belt of temperate forests carpeted what is now China, providing an ideal environment in which life could diversify. Fortunately for us, but not for the Jurassic residents, one of these forests was particularly close to a volcano. Occasionally, eruptions from this volcano covered the area in ash, helping to preserve as fossils the lifeforms that lived there. The beds of rock formed by these ashfalls, along with huge lakes, are called the Tiaojishan Formation, and they contain some of the most detailed and beautifully preserved fossils on Earth.

The Tiaojishan rocks are a snapshot of a green Jurassic paradise. It was warm and humid, a dense jungle of conifers, cycads, ferns, horsetails, moss and ginkgoes. Flying above was a diverse

mixture of both the primitive and advanced types of pterosaurs, all filling specialist niches. Amongst these was a bizarre frog-faced species called *Sinomacrops,* which had a wingspan of around 30 centimetres. This animal would have resembled a bat in many respects, using its large eyes and wide wings to hunt insects at dusk. Environments like forests can accelerate evolution because they are relatively productive, and it is difficult for any one species to dominate such a complex habitat. As well as pterosaurs, the Tiaojishan was home to a rich diversity of amphibians, lizards, spiders and insects. However, these exceptional fossil beds are most famous for their dinosaurs and mammals.

EARLY MAMMALS

There are three groups of mammals alive today: the monotremes, marsupials and eutherians. Of these, the monotremes are the most primitive, and today this group includes the platypus and various echidnas. Much like reptiles, these mammals lay eggs and have a single opening — the cloaca — which is used for reproduction, defecation and urination. The eggs are retained within the mother for some time before being laid. Platypus lay between one and three eggs in underground nests, with the mother curled around to incubate them. Echidnas go one step further and lay a single egg directly into a special pouch, which the mother has on her underside. When the echidna eggs hatch, the very small and underdeveloped young, known as a 'puggle', begins grappling to find milk using its strong front limbs. The mother feeds the puggle with milk released from openings in her skin rather than nipples. As the baby's spines begin to grow, they are released into an underground burrow where the mother will care for it for around a year. In monotremes, the bulk of the nutrient transfer and growth happens outside the mother's body. Marsupials and eutherians keep the embryos within their bodies, rather than laying eggs. In marsupials this may only last a short while before the embryo leaves the womb to develop in the mother's pouch, but eutherians retain their young for much longer, and by investing more in the developing young when they are born, the chances of survival are much greater.

The earliest eutherian in the fossil record is *Juramaia,* a 10-centimetre-long, shrew-like creature from the Tiaojishan Formation. *Juramaia* represents a huge leap in parental care; from tiny creatures like this, all eutherians would evolve — around 95 per cent of the mammal species alive today. While *Juramaia* may have superficially resembled a shrew, the two mammals actually evolved over 100 million years apart. There were other lookalikes lurking in the Jurassic

forests, too. Several rodent-like species lived on the ground, and one even resembled a mole, presumably burrowing in the same way. Convergent evolution had led to two different lineages solving a problem in the same way, and above the forest floor some pioneering mammals were tackling a very grand problem indeed.

JURASSIC GLIDERS

One of the biggest challenges of living in a forest is using the vertical spaces and accessing food higher up in trees. Climbing is one option, but travelling from one tree to another requires going all the way down to ground level and up again. To make this easier, save energy, and escape from predators quickly, many animals today have evolved gliding. The flying dragon lizard (*Draco*) of Southeast Asia is an excellent climber, but it can also glide using its winglike skin extensions.

ABOVE: A juvenile *Sinraptor* approaches the edge of a canyon in the Late Jurassic forests of what is now China. This carnivore was a distant relative of *Allosaurus* and other more primitive theropods.

These are actually modified ribs, which the lizard can use to travel tens of metres at a time. Creating a larger surface area to resist air is a common strategy of gliders, although in many cases it is more accurate to describe it as falling with style. Examples include the gliding geckos, as well as frogs that have enlarged foot pads and skin flaps on their body in order to do this. Surprisingly, some of the best gliders are the 'flying snakes', which slither through the air while pulling in their bellies. By making their undersides concave, they trap air like a frisbee, allowing them to travel an incredible 100 metres.

A number of modern mammals are also capable of gliding, including flying squirrels and gliding possums. These use flaps of skin that are stretched between their extended limbs, creating a large surface area, rather like sails, to catch the air. In the Tioajishan forests there were a number of very similar-looking mammals that could probably also glide in this way. They possessed a membrane between the limbs, and would have been very effective climbers, much like flying squirrels and sugar gliders. Again, despite the striking similarity, these Jurassic gliders were only very distantly related to anything alive today. Even just a little extra area of skin between the limbs would have given these animals the edge when jumping from tree to tree. Over time, the worst gliders would be picked off by predators or hunger, and the most successful would thrive in this vertical world.

TAKING TO THE WING

By this time, another group of theropod dinosaurs of the coelurosaur group was beginning to tackle gliding as well – they were the early ancestors of birds. Some, like the tree-dwelling species *Yi qi,* were taking a similar approach to flying squirrels. As well as simple down-like feathers, *Yi qi* had a membrane of skin between its arms and body, forming a kind of wing. This membrane was supported by a long finger, which would have given it a strange, bat-like appearance in the air. This was ultimately an evolutionary dead end, and creatures like *Yi qi* eventually disappeared. The bird ancestors were beginning to develop much more advanced wings, and the key was their feathers. All the Chinese theropods had feathers, but these were usually nothing more than single or simply branched filaments. These would have appeared fluffy or shaggy on the animal, and would have provided thermal insulation in much the same way as mammal hair. However, bird ancestors began evolving feathers that were branched with a stiffer central axis, giving them a distinct shape and rigidity. Arranged along the arms, they formed a strong wing, increasing the surface area for lift enormously without much added weight. Having

LEFT: *Anchiornis* was one of the earliest ancestors of true birds, evolving in the Late Jurassic forests of China. It possessed primitive feathers that were used for display and gliding.

NEXT PAGE: By studying microscopic structures within feathers, scientists have been able to map the colours of *Anchiornis* and other exceptionally well-fossilised dinosaurs.

a light wing like this would have made the bird ancestors of the Tiaojishan quite accomplished gliders. Among them was *Anchiornis,* a small 40-centimetre-long theropod, which had shaggy feathers over most of the body, but well-formed feathers on its wings, legs, tail and even its head. Remarkably, the exceptional fossilisation of *Anchiornis* has enabled scientists to identify microscopic structures in the feathers that are associated with colour. From these studies, we know that *Anchiornis* was mainly grey, but it had a reddish head crest and specks of black and white on its more-developed feathers.

A crucial next step towards true flight was the ability to generate lift by flapping the wings, rather than just passively gliding. Powered flight has enormous benefits, such as being able to

BELOW: The Berlin specimen of *Archaeopteryx* is one of the most recognisable fossils ever discovered, and offered some of the first clues that dinosaurs and birds were related.

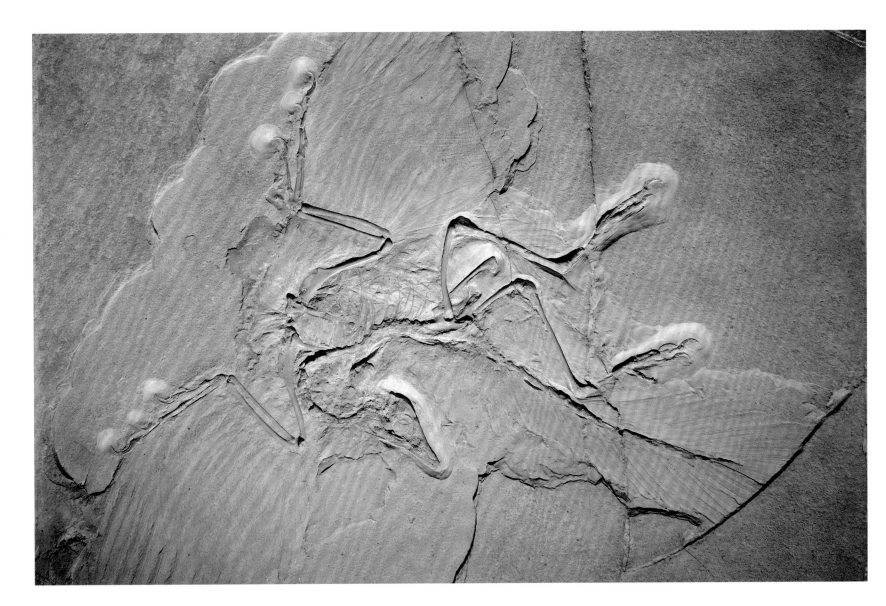

travel longer distances, escaping predators and exploiting airborne prey like insects. However, the exact reason for the evolving flapping flight is still debated. Interestingly, some birds today are capable of flapping their wings backwards, in such a way that they can push themselves downwards rather than upwards. Bird ancestors like *Anchiornis* had claws on their feet and may have used these in combination with backwards flapping to help them run up vertical surfaces like tree trunks. Normal flapping, to generate lift, may have begun to help these animals leap higher to catch flying insects, or perhaps just give them that extra little bit of distance while gliding. Whatever the reason, the exceptional fossils of the Tiaojishan are a window into the evolutionary experiments that led to powered flight in birds. They helped scientists refine the idea that birds were descended from theropod dinosaurs, and, in fact, the 160-million-year-old *Anchiornis* and its relatives may represent the earliest-known birds. However, for a long time this title belonged to a much more famous flying dinosaur, the *Archaeopteryx*.

Archaeopteryx lived about 10 million years after *Anchiornis*, during the Late Jurassic in what is now Germany. At the time, high sea levels had flooded Europe, forming archipelagos dotted with peaceful lagoons. Some of these lagoons would periodically evaporate, before being refilled with seawater, and as this happened they became much saltier and unable to support life. The rocks that formed at the bottom of these gentle, yet deadly, lagoons are very fine-grained, buff-coloured and split into beautifully clean plates like the pages of a book. In fact, these plates are so flat and the rock so pure that they were originally quarried for use in early printing. Fossils preserved in these rocks are equally beautiful, and they often provide snapshots of the last moments of the animal's life. One morbid discovery was of a fossilised trackway made by a horseshoe crab that had fallen into the toxic bottom waters. Unable to breathe, this unfortunate creature died and became perfectly preserved, leaving an almost 10-metre-long death track behind it. In another extraordinary case, a large predatory fish has been fossilised with a pterosaur in its jaws, apparently choking on it before sinking to the bottom of the lagoon. What is remarkable is that the pterosaur victim itself had a smaller fish in its throat. So the smallest fish had been hunted by the pterosaur shortly before the pterosaur had been snatched by the bigger predator, which died trying to swallow both.

These deposits provide wonderful insight into the diversity of the tropical Late Jurassic, home to an abundance of fishes, turtles, ichthyosaurs, crocodile ancestors, pterosaurs and more. Fossils of the most famous resident – *Archaeopteryx* – have been excavated since 1861, shortly after the publication of Charles Darwin's seminal work *On the Origin of Species*. It had clawed fingers, pointy teeth and a long tail, but it also had feathers and a bird-like skull. As a result the fossils were either identified as birds or theropod dinosaurs. Eventually, it was recognised that *Archaeopteryx* was a 'missing link' between the two, and much later discoveries of feathered dinosaurs in China would further confirm this idea. We now know that dinosaurs had a lot of the required anatomy for flight, well before its evolution. As far back as the Triassic, feathers had evolved, as well as hollow bones that made the skeleton much lighter. The flying abilities of *Archaeopteryx* have been debated for decades; however, it is likely that it was capable of powered flight, albeit with less strength and flexibility than modern birds. These kinds of bird-line theropods had, throughout the Jurassic, become much smaller and well-adapted to life in dense and complex environments. This was in sharp contrast to other allosaur- and megalosaur-line theropods, which had taken quite a different evolutionary path.

GIANTS OF THE END-JURASSIC

By the Late Jurassic Morrison Formation of North America, some truly enormous dinosaurs had evolved, with predators and prey locked in an arms race to outsize each other. The 1.5-million-square-kilometre expanse of prairie, forests and floodplains was home to some of the most iconic dinosaur species, and one focus of the 'Bone Wars' conflict of the late 1800s. Theropods, like *Allosaurus,* were common and fearsome predators of these lands, and at well over 9 metres long they were capable of taking on almost anything. Studies of their skulls have shown that they probably weren't very intelligent but their sense of smell was excellent. The skulls also reveal that when *Allosaurus* attacked it opened its jaws very wide and used an axe-like strike to inflict injury. This attack behaviour and a mouth full of sharp, serrated teeth would have been effective against large prey, such as the 7-metre-long ornithischian *Stegosaurus*. *Allosaurus* also lived alongside a number of gigantic sauropods, such as the 20-metre-long *Brachiosaurus* and 25-metre-long *Diplodocus*. Both of these dinosaurs would have required vast quantities of vegetation to fuel their growth, especially in their early years as vulnerable juveniles. To avoid competition they likely targeted different plants, with *Diplodocus* feeding on low-lying vegetation and *Brachiosaurus* reaching up to feed on conifer leaves.

The Jurassic had been a time of warm temperatures and continental fragmentation, forming lush forests and shallow seas teeming with life. It was a time of recovery, diversification and innovation in the wake of a mass extinction. The period ended with almost double the diversity of life it had started with, but another climate hurdle was to come; the end of the Jurassic saw a dip in temperatures, a cold snap that would challenge even the hardiest polar animals.

RIGHT: The theropod *Allosaurus fragilis* of the Late Jurassic of North America was a large and quite common predator. Healed injuries on their bones suggest they lived particularly violent lives.

NEXT PAGE: Baby *Diplodocus* were tiny miniatures of their parents, left to fend for themselves after hatching. Their best defence was to hide and grow up as quickly as possible.

PARADISE LOST

The Cretaceous period began 145 million years ago. It was a time of global cooling, when snowfalls were frequent towards the poles, and there may even have been glaciers and ice sheets, especially in the mountains. Earth was the coldest it had been for over 100 million years, and the high latitudes were uninhabitable for any creature built for warmer temperatures. The sauropods, for example, which relied on consuming large quantities of vegetation, were restricted to the lower latitudes where the forests were more productive. By this time Gondwana had broken up, leaving Antarctica, Australia and India lumped to the east, and Africa and South America to the west. This dismantling of Pangaea continued to isolate populations of land animals, even more so than in the Jurassic period. Without the conflicts of a world without barriers, smaller pockets of animals and plants were free to diversify in their own unique ways. Soon after the dissolution of Gondwana, India and Madagascar began drifting northwards together, remaining isolated for the rest of the Cretaceous. During that time, a number of unique species would evolve there, in much the same way as Australia and New Zealand evolved their own distinct faunas.

After a few cold snaps in the Early Cretaceous, temperatures eventually returned to tropical Jurassic levels. The increased tectonic activity that was fragmenting the

PREVIOUS PAGE: Following the devastating impact of an asteroid, Azhdarchid pterosaurs try to escape the searing heat of a burning North American forest.

RIGHT: Conifers were well adapted to cold and dark winters at higher latitudes, a habitat they still dominate today.

world's continents was also fuelling volcanism and a rise in carbon dioxide. Global average temperatures fluctuated, but ultimately they rose over the next 50 million years, reaching a warm peak known as the 'Mid-Cretaceous Hothouse'. Throughout this time, continents were drifting apart but few were colliding, which meant mountains were not being formed. Usually, when continents collide the rocks buckle and compress, thrusting mountains upwards at the point of impact. As this happens, new rock is exposed to the air, and as it weathers it reacts with carbon dioxide, removing it from the atmosphere. Without mountains, the rise in global temperature would have created particularly potent conditions, and the rising sea level at this time meant that a third of the land's surface was covered with shallow seas. These warming seas were dominated by an extinct group of clams called rudists, which had conical shells that could grow to over a metre long. Throughout the Cretaceous, rudists became the major reef builders, growing and spreading quickly in high-temperature waters that would have been lethal to the corals.

THE FIRST BLOOM

The landscapes of the Jurassic had been dominated by seed-producing plants like cycads, conifers and ginkgos. These ancient plants are known as gymnosperms, which literally means 'naked-seed', and are so-called because their seeds are formed on the surface and are relatively exposed. In many cases the leaves are modified into protective structures like cones. Cycads and conifers, in particular, were very successful in the Jurassic and would have been an important food source for dinosaurs, especially the softer new growth. In the understory, ferns ruled the damp shadows, living much as they had done for over 200 million years. Ferns reproduce using spores, relying on water for the sperm to swim in and fertilise an egg. Gymnosperms, however, use pollen for fertilisation, a mass of tiny hard-coated grains, each of which are a factory for sperm. In gymnosperms, the pollen is released and spread, then activates when it comes into contact with a corresponding female structure. Cycads and ginkgoes have sperm with tails, which can actively swim from the pollen to the egg. Conifers have pollen, which produces a long tube that transports the sperm directly to the egg. Whatever the method of fertilisation, pollen was a huge innovation for these plants, as it protected the sperm-producing cells over longer distances. Gymnosperms evolved in the Palaeozoic, and for a long time would have relied almost entirely on the wind to spread their pollen. By the Triassic period, though, gymnosperms were spreading their pollen in more unusual ways; a fossilised cycad cone from Antarctica was found with tiny beetle faeces, which contained pollen grains smaller than the width of a human hair that belonged to the cycad. These Triassic beetles were eating the pollen and would have been carrying it on their bodies between cycad plants, making them amongst the earliest pollinators on Earth.

Enlisting the help of insects was a game-changing innovation, one that vastly improved the success rate of pollination. In the earliest stages of these kinds of relationships, the pollinators would have been unspecialised, going from one plant to another without much regard for the food source. Likewise, plants would have benefited greatly from this pollen transport, and would have been unfussy about who was visiting them. However, as more plants began using these insect couriers, competition grew for their services. The natural conclusion was to form closer

relationships, with insects and plants committed to each other. What resulted would become the most successful advertising campaign that the Earth had ever seen: the evolution of flowers. It is difficult to imagine a world without the smell, colour and exuberance of flowers.

Today, flowering plants – also known as angiosperms – are a hugely important part of modern ecosystems and make up a massive 80 per cent of living plant species. Almost everything we think of as a plant – from oaks to orchids, cacti to chamomile – is an angiosperm. Their sheer diversity is nothing short of astonishing. However, scientists are still unsure about exactly when the first angiosperms evolved. There are some intriguing Jurassic contenders, but identifying the very first of any group of organisms is fraught with difficulty.

The earliest uncontroversial angiosperm in the fossil record is *Archaefructus*, found in 125-million-year-old rocks of the Yixian Formation, in northeastern China. In life, it would have inhabited lakes, with shoots extending about 50 centimetres out of the water. Unlike modern angiosperms, it didn't have flowers with petals, but it did have the reproductive structures of a flower along the stem. The carpals are the female parts of the plant, and they hold the eggs, which become seeds. The male parts are the stamens, which produce and release pollen. Both of these structures are modified leaves, which have been adapted to perform reproductive duties more efficiently. *Archaefructus* literally means 'ancient fruit' because the seeds are surrounded by a protective coating, the fruit. Today, many fruits are packed with an assortment of sugars, vitamins and minerals that animals find attractive. They are often brightly coloured in a way that maximises their contrast with the surrounding leaves.

Fruit was an important innovation for plants, but the question remains, why make your seeds so appetising? While it may seem counterproductive, the benefits of encouraging animals to eat fruits far outweigh the cost of a few seeds being lost to digestion. Seeds have a tough coating that can survive digestive tracts, and while they are inside the animal they can be carried for many kilometres. When they are finally excreted, they are left sitting in a pile of fertiliser, far away from the parent plant. An added benefit is that any insect larvae that may otherwise eat the seeds are killed during this process, so the dung is sterile as well as nutritious.

Fruit-like structures have evolved in many groups of gymnosperms as well. Juniper 'berries', the main spice used for gin, are cones, but unlike pine cones they don't open up their woody scales to release their seeds. Instead, juniper cones become soft and fleshy, and rely on small animals to eat and disperse them. All major lineages of gymnosperms have members that produce some kind of appetising cover to their seeds, and these have evolved in slightly different ways. This demonstrates how effective this method of seed dispersal really is, and why plants go

LEFT (ABOVE):
A scarab beetle covered in pollen on the leaves of a giant water lily, having fed within it overnight.

LEFT (BELOW):
A fly visits a flower to feed on its nectar and picks up pollen in the process. This strategy aids fertilisation across larger distances, and has helped flowering plants evolve and spread quickly.

to such great lengths to produce fruit. It is such an important process that in some rainforests today 90 per cent of angiosperm trees use animals to disperse their seeds in this way.

The angiosperms' most famous and iconic innovation was the flower. Unlike *Archaefructus*, which had the reproductive structures along the stem, later angiosperms had these parts arranged in a compact structure. So instead, the carpals and stamen were surrounded by petals, modified leaves that would protect these structures and advertise them with bright colours when the time was right. These adverts are often targeted very specifically to certain pollinators; as well as the huge palette of colours that we humans can see, many flowers also have display patterns in the ultraviolet spectrum for insects.

A REVOLUTION BEGINS

By partnering with animals for reproduction and dispersal, angiosperms had a distinct advantage over other plants – they were capable of spreading and occupying habitat rapidly. Large dinosaurs moving through a forest would have caused huge amounts of damage, as would natural disasters like earthquakes or wildfires. In the wake of this kind of chaos, the first plants to claim the space would have been the fast-growing angiosperms, whose seeds could have been lying dormant on the forest floor for years. This process can be seen today in newly felled forest or unattended gardens; within days or weeks there will be annual flowers germinating, and within a few years angiosperm shrubs will have taken over. This ability of angiosperms to spread like weeds and adapt to changing environments would fundamentally alter the texture of Earth's landscapes forever. As angiosperms took over almost every habitat on the planet, their diversity exploded. Countless new species blossomed for the first time during the Cretaceous, and with every new angiosperm came opportunity for the animals with which they formed partnerships.

For the majority of life's 4-billion-year existence, the oceans had fostered more biodiversity than the land. However, since the Cambrian a slow but relentless march had been underway, as animals and plants made permanent homes above sea level. Today, despite covering over 70 per cent of the Earth's surface, the oceans now only contain 15 per cent of the millions of known species alive. At a crucial point in the mid-Cretaceous, for the first time in Earth's history, the number of species on land overtook those in the sea. From around 110 million years ago, this revolution saw an explosion of biodiversity, fuelled by the success of flowering plants. Insects, spiders, reptiles, birds, amphibians and mammals all diversified, including many of

RIGHT (ABOVE):
The Early Cretaceous theropod *Deinonychus* may have had more advanced social behaviour than other dinosaurs, including better parental care.

RIGHT (BELOW):
A juvenile *Deinonychus* shows off its long recurved killing claw, used like the talons of modern birds of prey. These were held off the ground while walking to keep them sharp.

the modern families that we would recognise today, such as snakes. This revolution lasted for over 50 million years, boosting the number of species increasingly higher as ever-more-complex ecologies developed. With new flowers came new pollinators, herbivores, fruit specialists, seed specialists and a suite of animals calling the plants home. And, of course, waiting in the wings were new predators for them all.

Insects were particularly important during this time because of the close relationships they developed with flowering plants. Older groups like true bugs, flies, beetles, termites and wasps all ballooned in diversity, and new groups such as ants and bees appeared. The success of ants and termites is particularly significant because of their role in maintaining the health of ecosystems, and their astonishing abundance. Today, the total weight of ants and termites may equal that of all other insects combined, and in the Amazon rainforest alone they make up a third of the animal biomass.

Ants evolved from stinging wasp ancestors around 100 million years ago, and while rare at first, they quickly occupied large expanses of Laurasia, the northern block of continents that included North America and Asia. They have since spread to almost every continent on Earth, living in colonies that can be millions strong. With 10,000 species, it is perhaps unsurprising that they have evolved a wonderful array of lifestyles and interactions with the animals, plants and fungi around them. There are farmers, engineers, parasites, and some even tend insects that feed them honeydew. The vast majority are predators or scavengers and can be important controls of the populations of other insects. Some ants protect plants that are beneficial to them, either as shelter or a food source. The lemon ant of South America, for example, can form colonies of 3 million workers, and these labour tirelessly to kill all but a select few of their favourite species of tree. These patches of fiercely curated forest have been termed 'devil's gardens', and a single colony of lemon ants can control up to 600 trees.

Of particular interest is the social structure within an ant colony, especially because of depictions of the 'hive mind' in art, and comparisons with our own civilisation. Ant colonies mostly consist of female workers and soldiers, which are sterile, and a small number of fertile queens and male drones, which can reproduce. As a group they communicate closely and operate as a collective, performing tasks much greater than the sum of their parts. As part of a superorganism, ants can overwhelm large prey, maintain huge territories and fundamentally alter the habitat around them. Survival of the fittest no longer applies to the individual, but to the entire colony. Part of the reason insects have been so successful is their wings, which allow

LEFT: Termite mounds have been recorded at over 12 metres high. They provide a great deal of protection against predators and the weather.

them to travel long distances for food and to spread across territories quickly. The wings are essentially dead tissue, which is incredibly lightweight but also easily damaged, and unlike the wings of flying vertebrates, they cannot be healed. The fragility of the wings limited what the individual could achieve in its lifetime, but by becoming truly social, ant colonies overcame this obstacle. The damage or death of one individual didn't hold back the army around it, or the genes that helped them survive. While most of the ants in the colony are wingless, there are special males and females with wings that leave the nest in spring or summer when it is hot and humid. The males usually leave first, secreting a pheromone that females in the area follow before the two mate. The fertilised female then finds a new nesting spot elsewhere, before laying eggs – the future workers of the colony.

Unlike ants, termites evolved from a branch of cockroaches that had become specialist wood-eaters. Before termites, the only organisms capable of breaking down the lignin in wood were fungi, but this was a very slow process. Termites used their powerful jaws in combination with microbes in their gut to digest huge quantities of dead and decaying wood relatively quickly. Old vegetation was cleared away much more rapidly following forest fires and storms, allowing aggressive and fast-growing plants

LEFT: Following a successful raid, *Megaponera* ants carrying termite prey back to their nest.

like angiosperms to thrive. Sociality in this group probably evolved in the Early Cretaceous and is similar to the division of labour seen in ants. Workers do most of the foraging and nest maintenance, but they also digest food for the other nestmates. Unlike ants, the workers can be male as well as female, sometimes with different jobs to perform. There is also a special king termite, which mates with a queen for life. She can produce 40,000 eggs a day when the colony matures and live for 50 years.

As ecosystem engineers there is natural competition for resources between ants and termites today, and this antagonism has likely been raging since the Cretaceous. In fact, ants are the main predators of termites, and some, such as the *Megaponera,* are specialist termite-eaters. Once a termite nest has been discovered by a scouting female *Megaponera,* a raid will begin, involving around 500 ants marching in a long column, two abreast. When they get close to the termite nest, they fan out and begin to surge forwards. Armed with large jaws the ants attack, but termites have a caste of armed soldiers to defend themselves. These soldiers can quickly catch and dismember ants, protecting the workers. The ants' best weapon is their venomous stings, which are effectively thrust between the jaws, where the termite is weakest. The ants will take a huge number of dead and dying termites back to their nest but will always leave some behind alive so that they can raid the nest again in future. Remarkably, injured ants who have lost one or two legs are retrieved from the battleground and taken care of by the other ants. The workers lick the wounds, using antibiotics carried in their saliva to help them heal, thus reducing the death rate of injured ants by 90 per cent. Survivors can help around the nest as usual and will even join future raids. Other than humans and chimpanzees, this is the only known example of animals using antimicrobials to help treat infected wounds.

Above the heads of the Cretaceous ants and termites, flying insects were also diversifying rapidly, and it is around this time that spiders began to construct aerial webs to catch them. Tiny filaments of silk web have been found trapped in amber – fossilised tree resin – along with the insects' prey. Silk had been used by spiders ever since the Devonian to reinforce their burrows and protect their eggs, but slowly they had begun to use it for triplines and nets on the ground. The silk itself is a focus of scientific interest because of its incredible strength – ten times that of kevlar. Webs of the Darwin's bark spider, for example, can span 25 metres, despite it only measuring around 2 centimetres across. Around 110 million years ago, the bounty of insects in the air saw many spiders scaling new heights, constructing elaborate and fascinating webs. However, the spiders were far from alone in catching these newly evolved aerial insects, because birds were well on their way to mastering powered flight.

RIGHT: By the middle Cretaceous, angiosperms were providing complex habitats in which invertebrates could diversify, including predators like web-spinning spiders.

DINOSAURS OF THE EARLY CRETACEOUS

Since *Archaeopteryx*, birds had continued to reduce the weight of their bodies for flight, at first losing their long and heavy tail. The function of the tail had been to increase surface area for lift and provide balance while flying, but this had now been replaced by a lighter array of feathers. Early Cretaceous birds, such as *Confuciusornis* of China, had a greatly reduced tail and had evolved a breastbone to support flight muscles. It was also the first bird with a beak, with a hard fingernail-like material covering the front of the skull. This beak was clearly very effective for feeding because within the jaws *Confuciusornis* did not have any teeth. Several hundred fossils of this crow-sized creature have been excavated, often with complete skeletons and traces of feathers left behind in incredible detail. Intriguingly, some of the fossilised birds have two long feathers on the tail, which may have belonged to male birds using them to attract females. Elaborate plumage is commonly used by modern birds to demonstrate their physical fitness to the opposite sex, and there is evidence of many ancient birds doing the same thing. Alongside *Confuciusornis*, there were around 40 different birds of various kinds living in the same habitats, many still with teeth and clawed wings. Throughout the Cretaceous, birds would begin to use their legs more for grasping, and lose their teeth in favour of a beak.

The 133- to 120 million-year-old rocks that preserve this incredible detail are collectively called the Jehol, and they lie above the Tiaojishan Formation that we explored in the last chapter. Like the earlier Tiaojishan beds, which contained a snapshot of the Jurassic ecosystem, the Jehol is a wonderful showcase of Early Cretaceous creatures. As well as birds, the Jehol preserves a number of other dinosaur groups which diversified in Asia at this time, and which would spread all over the world by the end of the Cretaceous. Among these were very early members of an ornithischian group, called the ceratopsians, which would one day include the iconic *Triceratops*. Unlike their descendants, the Jehol ceratopsians were small and ran on two legs. Curiously, they had quill-like feathers extending from the tail like a porcupine, although their exact function is a mystery. The Jehol was also home to early members of a strange group of theropods called the therizinosaurs, close relatives of the bird-line theropods, which had elongated claws on their arms. The claws were the first part of the animal that was discovered, and were initially interpreted as a weapon. However, when a skull was found, it became clear that this dinosaur was a herbivore, with teeth perfectly adapted for grinding up leaves. So the claws were probably used to pull down branches to feed, which was used to great effect in combination with its long neck and pot belly full of fermenting bacteria.

RIGHT: The resplendent quetzal lives in dense forest, and so its feathers, like those of other forest birds, are often brightly coloured for communication and sexual displays.

Perhaps the most famous residents of the Jehol were the tyrannosaur relatives, a number of which stalked the wetlands and forests of this ecosystem, and which had evolved in Laurasia and probably never made it to Gondwana. Some of the earliest tyrannosaurs to evolve in the Jurassic were small, lightly built, and had three fingers, comparable to other non-bird theropods of the time. However, throughout their evolution they became much larger and more robust, and some from the Jehol were reaching 9 to 10 metres in length. This cemented their place as the top predators of Laurasia, and in time these kinds of dinosaurs would outcompete the older megalosaur- and allosaur-type species.

Tyrannosaurs were filling the larger predator niches all over Laurasia throughout the Cretaceous, and at their feet other theropods were tackling the smaller prey. Bigger and badder enables predators to tackle the largest trophies, but it also limits them to hunting in open and fairly flat landscapes. There will always be a place for the small and agile hunters, which are able to explore the dense forests, steep hillsides and other complex habitats that the giants cannot access. In the Jehol and other parts of the world, these niches were filled by the dromaeosaurs, another group closely related to the bird-line theropods. Dromaeosaurs had first appeared in the Middle Jurassic and had spread worldwide by the Cretaceous – except in India, which was still an isolated island continent. China may have been isolated through much of the Jurassic, too, but by the Cretaceous land bridges were allowing animals and plants to spread.

As well as accessing more unusual habitats for prey, the dromaeosaurs had other advantages over their tyrannical big cousins. Dromaeosaurs had an unusually large claw on each foot, which was curved and sharply pointed. Originally it was thought that this claw was used for slashing, but more recent studies suggest it was used for piercing and holding onto prey, like a meat hook. These claws were held off the ground while walking to keep them sharp, and would have been effective for pinning down animals – just as birds of prey do today. Once captured, the dromaeosaurs would butcher the kill by biting and pulling its head backwards with a mouth full of sharp, curved teeth. These dromaeosaurs were fully feathered and retained their 'wings', so it may be that the body was stabilised by flapping as they attacked their prey, giving these animals almost gymnastic agility.

The Early Cretaceous of North America was home to a particularly well-endowed dromaeosaur, *Deinonychus*, whose name literally means 'terrible claw'. This dinosaur was discovered in 1931 by palaeontologist Barnum Brown and his team, who around 30 years earlier had discovered the first remains of *Tyrannosaurus rex*. Teeth from several *Deinonychus* individuals have been found scattered around the body of a large ornithischian dinosaur called *Tenontosaurus*, which they may have been feeding on as a group. It would certainly give the group an advantage to hunt in packs, and it would have allowed these 3-metre-long predators to hunt larger prey.

Deinonychus lived around 110 million years ago, in what was then a floodplain, alongside a rich diversity of ornithischian dinosaurs, sauropods and other theropods. One of these was *Arkansaurus*, a theropod belonging to a group called the ornithomimids, or 'bird-mimics'. Despite the group being quite closely related to *Deinonychus*, they looked more like ostriches – fully feathered and standing tall with a long neck and a small, narrow head. Ornithomimids like this were the most common small dinosaurs in North America, and they would have been a frequent prey item for anything capable of catching them. Throughout the Cretaceous these animals gradually evolved a beak and lost their teeth, suggesting they were herbivores.

The Cretaceous had begun with the lowest global temperatures for 100 million years, but a lot had changed by the middle of the period. Tectonic activity had accelerated, and as continents fragmented, lava flooded to the surface, up through the fissures created between them. As more

BELOW: A group of *Deinoncyhus* stalk prey in the Early Cretaceous scrublands of North America.

continents pulled apart into ever-smaller pieces, the number of these fissures increased as well, and with the lava came carbon dioxide.

Global warming ensued, driving up the heat to its highest level since the Permian–Triassic mass extinction had almost wiped out life on Earth. It reached a peak 94 million years ago, with a global average temperature of 34°C, and a whopping 13°C at the poles. If you've read the whole story this far, you know how history likes to repeat itself, and what came next. With the heat and carbon dioxide came 500,000 years of ocean acidity and stagnation. Around a quarter of marine invertebrates were lost, and with them a number of iconic marine reptiles. Ichthyosaurs, those dolphin-like reptiles that evolved in the Triassic, had survived the end-Triassic mass extinction and the climate upheavals of the Jurassic. By the Early Cretaceous they were doing well, possibly because the division of continents and rise in sea level had created new shallow-water and coastal habitats. Despite their incredible success, and 160 million years of survival, the last ichthyosaurs fell at this huge climate hurdle. Another victim of the Cretaceous Hothouse was the pliosaurs, those paddle-finned plesiosaurs with short necks and huge crocodile-like mouths. Two of the greatest dynasties of marine reptiles were gone, and this left a power vacuum in the seas.

NORTH AMERICAN MONSTERS

Shortly before the extinction of these marine reptiles, a group of small lizards known as mosasaurs was taking to the water in what is now Europe. At this time, Europe was a series of islands and tropical shallow seas, a perfect hunting ground for these animals. Eventually, they evolved paddles, much like the plesiosaurs', and a long streamlined body to move through the water more easily. By 92 million years ago they had reached a metre in length, and some species had managed to cross the Atlantic Ocean. By the end of the Cretaceous, mosasaurs had spread worldwide, with some growing to over 15 metres. In the blink of an eye, mosasaurs had become apex predators, rivalling the last of the plesiosaurs and even the ancient sharks.

Since the Early Jurassic, sharks had been approaching their familiar modern forms, and in many cases equalling the ferocity of the massive marine reptiles. Direct evidence of this was found on a fossilised chunk of mosasaur tail that had shark bite marks on it. It is not uncommon to find bite marks on bones, which are often the work of scavengers who have been stripping the carcass, but this bone had signs of infection after it had been bitten: smoking-gun proof that these animals interacted violently. Some Late Cretaceous sharks reached 8 metres in length, too, larger than modern great white sharks; these were dangerous oceans.

One of the best-studied battlegrounds of Late Cretaceous predators is the Western Interior Seaway, which split North America in two: Appalachia lay to the east, Laramidia to the west. This stretch of sea was huge, over 3,000 kilometres long but very shallow at around 800 metres deep. Its sun-bathed tropical waters were especially productive and were teeming with life – joining the marine reptiles and sharks were coiled ammonites, squid-like belemnites and bony fishes, including some predatory species that reached 5 metres long. Despite the danger, pterosaurs were abundant here and even a few dinosaurs made a living out on the waves. The early birds *Hesperornis* and *Ichthyornis* were well-suited to catching fish, and both were armed with a beak and teeth. *Hesperornis* was 1.8 metres in length and flightless and would have resembled a giant grebe or loon, and it had powerful legs for diving and swimming and would have looked quite awkward on land. *Ichthyornis* was smaller, at about the size of a pigeon, and would have behaved more like a gull or tern, hunting fish from above.

ABOVE: Lamnid sharks, the group which includes the modern great white shark, first evolved in the Late Cretaceous alongside giant marine reptiles.

On land, the Late Cretaceous of North America was home to great herds of ornithischian dinosaurs called hadrosaurs, also known as the 'duck-bills' because of their wide snouts. These dinosaurs had evolved a way of chewing that involved complex movement of the skull and jaw, which remains completely unique in the animal kingdom. They also had batteries of small teeth that formed a large grinding surface, making them very efficient herbivores. This innovative feeding adaptation no doubt contributed to the hadrosaurs' success, and some, like *Maiasaura*, could form herds of thousands of animals. This would have been an impressive spectacle, with adults growing to around 9 metres in length and weighing 4 tonnes. The name *Maiasaura* means 'good mother reptile', which was given to them because of the discovery of a nest with eggshells and juveniles that were too big to be hatchlings. This meant that juveniles were being fed at the nest by the mother, and since this initial find, a great deal more has come to light about the life history of these animals. We now know that nests were constructed around 7 metres apart in large colonies of around 40 mothers, and that the juveniles would have vegetation brought to them by the mother until they were around a metre long. Some scientists have even suggested that this diet may have been supplemented by the mothers producing a kind of milk. While this may seem odd, pigeons, doves, emperor penguins, flamingos and many seabirds produce a milk-like fluid or oil for their young. The composition of these milks is extraordinary, with flamingo milk containing 54 times more protein than that from humans, and pigeon milk helping their young grow three times faster than other birds. Currently, this is just healthy speculation, but it may explain how *Maiasaura* juveniles survived on regurgitated fodder plants such as gymnosperms, cycads and ferns, which only have around 60 per cent of the energy of modern grasses.

Maiasaura lived about 77 million years ago on the western shoreline of the interior seaway, alongside other hadrosaurs, tyrannosaurs, dromaeosaurs, ceratopsians and ankylosaurs. The ceratopsians had batteries of grinding teeth similar to those of the hadrosaurs and became abundant by the Late Cretaceous. Perhaps in response to the predatory theropods, ceratopsians had evolved impressive facial horns and head shields, focusing their armour on vulnerable parts of the body like the neck. Many species were also getting much bigger as a form of defence, with some ceratopsians reaching up to 6 metres. Tyrannosaurs, at the time, were approaching 9 metres high, so it may be that the ceratopsians were also herding for extra protection.

By far the best armoured dinosaurs of the Late Cretaceous were the ankylosaurs, which were built like tanks, walking on all fours and feeding close to the ground. They had skin peppered with bony plates known as osteoderms, and robust spikes and other projections all over the body for protection, as well as a huge club tail that it could swing at predators, or perhaps

LEFT (ABOVE):
A herd of *Maiasaura* guard their young. Discoveries of these ornithischians alongside their nests have provided evidence of complex parental care in dinosaurs.

LEFT (BELOW):
Maiasaura nests were constructed from mud and were lined with vegetation, which provided heat for incubation as it rotted.

NEXT PAGE:
Maiasaura and other hadrosaurs may have travelled in huge herds of thousands of animals. During their migrations, they were seeking food and water.

rival ankylosaurs. *Maiasaura* and its other dinosaur contemporaries lived in what was then a semi-arid region of North America, which at the time experienced long dry seasons and warm temperatures. However, Late Cretaceous dinosaurs were by no means limited to these habitats.

ARCTIC DINOSAURS

Although temperatures were subsiding in the later Cretaceous, they were still quite high by modern standards. Despite lying close to the North Pole, Alaska was around 6°C on average and had no permanent ice, compared to a temperature of -12°C today. Deciduous trees, ferns and horsetails all survived here, despite enduring four months of darkness and markedly colder temperatures in the winter. Dinosaurs survived here as well, including horned ceratopsians and massive hadrosaurs like *Edmontosaurus*. These ornithischians could reach 12 metres long, and the better part of 6 tonnes, but despite this, may have been able to reach speeds of over 50 km/h when galloping.

The question of how these enormous creatures fed themselves has led scientists to explore whether this and other hadrosaurs may have migrated. As dinosaurs grow, they lay down bone, and this can reflect the quality of what they are eating. The microscopic structure of hadrosaur bone shows how the animals grew compared to the southern populations. When food is scarce, bone is laid down more slowly and looks like the rings of a tree, but when food is abundant, it looks like a sponge of speckles and blobs.

Unlike the hadrosaurs living in warmer temperatures, the Alaskan dinosaurs had regular bands of these two bone types, reflecting poor food availability in winter and a bounty of food and growth in summer. These animals were putting up with the cold and dark for months on end, but despite these challenges *Edmontosaurus* and other herbivores were quite abundant. Recent studies have confirmed that a great many dinosaurs were not only overwintering but also breeding in this habitat in the summer months. Of course, with lots of prey come predators adapted to take advantage, even in the harshest environments. Theropods such as the 4-metre-long theropod *Troodon* had huge eyes, to make the most of the endless winter night, and probably had thick downy feathers to keep it warm. There was even a tyrannosaur, called *Nanuqsaurus*, named after the Native Alaskan Iñupiat word for 'polar bear'. While *Troodons* living towards the North Pole were notably bigger than their southern cousins, *Nanuqsaurus*, at 8 to 9 metres, is smaller than tyrannosaurs in the south. It appears that there may have been a Goldilocks size for predators in this environment – too big and they risk starving because they can't support themselves; too small and they risk starving because they can't hunt the larger prey.

RIGHT: Two mature *Tyrannosaurus rex* perform a courtship ritual in the Late Cretaceous of what is now North America.

NEXT PAGE: *Edmontosaurus* were among the largest herbivores of the Hell Creek Formation. Herds would have presented a dangerous challenge, even for *T. rex*.

THE TYRANT KING

The most famous tyrannosaur is without a doubt *Tyrannosaurus rex*, whose name means 'king of the tyrant reptiles' – and with good reason. At over 12 metres long and weighing the better part of 9 tonnes, this theropod has earned its rightful place as the king of the Late Cretaceous. *T. rex* is probably the most-studied extinct organism in history, and we know an unusually large amount about this animal.

Starting at the head, we know from the eye placement and brain that *T. rex* had keen vision, perhaps thirteen times better than ours, even surpassing that of modern birds of prey. Their sense of smell was also very strong, far better than most other theropods, and like modern vultures *T. rex* may have been able to smell a carcass from long distances. Its arsenal of heightened senses also included a snout that was very sensitive to touch, and hearing that was especially acute in the lowest frequencies. The skull was very large and robust for the animal's size, perfectly suited

BELOW: A mature male *Tyrannosaurus rex*, with long filamentous display feathers on his head. Evidence now shows that their teeth were completely covered in lip tissue.

to crushing bone. In fact, the bite force has been estimated to be the largest of any land animal, around 6 tonnes of force — the equivalent of biting down with the weight of a large African elephant. This force was transmitted through sharp banana-shaped teeth which could reach over 30 centimetres long including the root. This would have helped *T. rex* hold and dispatch its prey and make it easier to crush bone to reach the nutritious marrow inside.

The head was so massive and effective that the other killing tools typical of a theropod were not needed. The legs were stout and functional, perfect for carrying the weight of this enormous animal, but it was certainly not leaping onto prey. The feet had large robust claws for grip on the ground, but these were incapable of slashing or piercing like those of *Deinonychus*. Most incredible are the arms, which are tiny compared to the head, at just a metre long. *Tyrannosaurus rex* is certainly not alone in its shortcomings, and several other unrelated groups of theropods have similarly reduced front limbs. One of these is *Meraxes*, an almost 11-metre-long theropod described in 2022 and named after a dragon from George R.R. Martin's novel *A Song of Ice and Fire*. This dinosaur was very similar in form to *Tyrannosaurus*, despite only being distantly related. Again, this seems to be a case of convergent evolution, where the loss of the large front limbs is of common benefit. The researchers who described *Meraxes* suggested that the role of grabbing prey was slowly replaced by the jaws in these dinosaurs, and over time the front limbs became useful for lower-strength jobs like getting up off the ground. Some have even suggested that longer arms would have been a hindrance during hunting, getting in the way and easily broken during violent encounters.

Despite this impressive killing weaponry, adult *T. rex* are thought to have had relatively slow running speeds. Most studies estimate their adult walking speed was around 5km/h and a maximum running speed was around 30km/h. This is far slower than their prey could run, with the hadrosaur *Edmontosaurus* possibly reaching 50km/h at full tilt. This has led to an interesting idea about division of labour, and the possibility that they may have hunted in packs. While the adults were large, robust and fairly slow, the youngsters were lean and graceful. In fact, the juvenile *Tyrannosaurus* look so different from the adults that some scientists have even suggested they are a different species. Studies of the bones revealed that *T. rex* had this lightweight and agile skeleton up until about the age of 14, at which point they underwent quite a dramatic growth spurt. Between the ages of 14 and 18, these animals put on at least 600 kilograms of weight a year, possibly double that number at peak growth. During this rapid growth phase, the skeleton transformed into the robust slower-moving adult form, living to around 30 years old. If the juveniles had the speed, and adults had the power and the experience, it is possible they worked together in family groups to hunt more effectively. Although there is no direct evidence of this for *Tyrannosaurus* specifically, there are tantalising clues from other closely related tyrannosaurs. Skeletons of an adult, a teenager and two juvenile tyrannosaurs have been found fossilised together, and preserved trackways show three animals walking with each other. Counter to this idea is a 13-year-old specimen called 'Jane', which has the teeth marks of another *T. rex* on its snout. So it may be that these animals had loose coalitions, without the kind of advanced social structure we associate with mammal pack hunters. There is also a long-running debate about whether *T. rex* was a hunter at all, or rather just a scavenger. All sorts of lines of evidence have been cast in favour of one side or the other, despite the fact that most hunters alive today will happily scavenge if the opportunity arises.

HELL CREEK

The largest and most complete *Tyrannosaurus rex* skeletons are found in a succession of northwest American rocks known as the Hell Creek Formation, dating to between 68 to 66 million years old. The formation offers a wonderful snapshot of Late Cretaceous life, which would have seemed remarkably familiar to us today. The rocks here preserve a floodplain at the edge of Laramidia, the western landmass of the Western Interior Seaway. The forests and swamps were home to small mammals, crocodiles, frogs, lizards, snakes and turtles. In the sea, the marine reptiles and sharks still ruled, including mosasaurs reaching well over 10 metres long. Pterosaurs with equally huge wingspans soared above, the last of their kind, but were incredibly successful worldwide. Flowering plants such as palms and deciduous trees were finally matching the dominance of the conifers, ferns and cycads. Dinosaurs included ankylosaurs, ornithomimids, small theropods, the hadrosaur *Edmontosauru*s and strange dome-headed ornithischians called pachycephalosaurs.

However, the most abundant dinosaurs were the ceratopsians, particularly *Triceratops,* which made up around 40 per cent of the large Hell Creek dinosaurs. It is possibly one of the most iconic dinosaurs that has been identified – besides *Tyrannosaurs rex,* which by comparison made up around 25 per cent of the Hell Creek's large dinosaur population. At 8 to 9 metres *Triceratops* would have been formidable quarry for even the largest *T. rex*, but there is good evidence that the two species came into contact. For example, healed *T. rex* bite marks have been found on a *Triceratops* skull, proving that it survived the encounter. Unhealed bite marks have also been found on carcasses, suggesting *T. rex* did manage to kill these animals, or at least scavenged their bodies. These grand dinosaur enemies were clearly locked in an arms race, which had resulted

in extreme size and aggression. The Hell Creek has always captured the public's imagination because it tells the stories of dinosaurs like this. However, these rocks also record their demise and the end of an era that had lasted over 160 million years. Right at the top of the Hell Creek are the last sediments laid down in the Cretaceous period: a 66-million-year-old moment in history that saw the loss of 75 per cent of all species on Earth.

OBLIVION

It is fair to say that most major extinctions are caused by volcanism and their effects on climate, and for many scientists the end-Cretaceous extinction was no different. At the beginning of the Cretaceous, India had split away from eastern Gondwana and had been quickly drifting northwards, at about 20 centimetres per year. During its long journey, at around 66 million years ago, the island continent began passing over a particularly hot region of magma in what is now known as the

ABOVE: *Triceratops* were very common in the Hell Creek habitat of North America, and were the target of hungry *Tyrannosaurus*.

NEXT PAGE: An adult female *Tyrannosaurus* hunts with her teenage offspring. These youngsters could achieve greater running speeds, but lacked the deadly bite force of their mother.

Indian Ocean. This geological blowtorch, known as the Réunion hotspot, began melting the Earth's crust between India and Africa, and soon volcanic eruptions were raging. More than a million cubic kilometres of lava flooded to the surface just as the extinction was occurring. In previous chapters we have explored the devastating effects that such an eruption can have on climate, and so usually this would have been a straight smoking-gun cause of the extinction. However, another major event coincided with the extinction, which is difficult to ignore.

An asteroid the size of Mount Everest, travelling 20 times faster than a bullet, hit the Yucatán Peninsula of Mexico, with the power of 10 billion Second World War atomic bombs. It penetrated the Earth to a depth of 20 kilometres, as rock around it vapourised, blasting away in all directions – including into space. Magnitude-12 earthquakes instantly began travelling away from the impact. Within the first 10 seconds, the vapourised rock and asteroid reached such high temperatures and pressures that they didn't even produce light, although soon a white glow began to build. The plume of ejected material grew, and a blinding white heat illuminated everything within at least 1,000 kilometres. Anything that could see the plume would have been obliterated by the intense heat and carbonised in seconds. Within the crater the Earth's crust rippled like a raindrop hitting a pond, acting more like a fluid than solid rock. Over 10 minutes, the rock settled into massive 180-kilometre-diameter rings, marking the crater.

The global fallout was devastating. Category 9 to 10 earthquakes conducted through the Earth's crust, flooding any areas next to water with giant waves. The dark curtain of ejected material spread away from the crater at around 2 kilometres every second, towering 70 kilometres high and blacking out the skies. As the material that was blasted upwards began to fall back to Earth, it created friction with the air, and also heat. Particles big and small descending together caused the black skies to glow with thermal radiation, producing ground temperatures of 325°C, with a peak of 545°C. As the rock that was vaporised by the asteroid began to cool, it condensed like hailstones into tiny millimetre-sized spheres. On the ground, the air would have been thick with these spheres raining down, as the intense heat raged from the clouds above. Around the world, wildfires were sparked, destroying vast swathes of forest. As the clouds above cooled, the Earth was plunged into total darkness, the sky turned black from soot and sulphur-rich dust. Acid rain, wildfires and earthquakes continued for months after the impact. Without sunlight, temperatures plummeted by 25°C and, for at least six months, plants could not photosynthesise, destroying food webs from the bottom up. It was a long and devastating winter, one that saw the end of the non-bird dinosaurs, pterosaurs, ammonites, mosasaurs, plesiosaurs and many others. Birds and mammals were reduced to a fraction of their diversity.

RIGHT: An asteroid collides with Earth 66 million years ago, bringing to an end the Cretaceous period, and the reign of the giant dinosaurs.

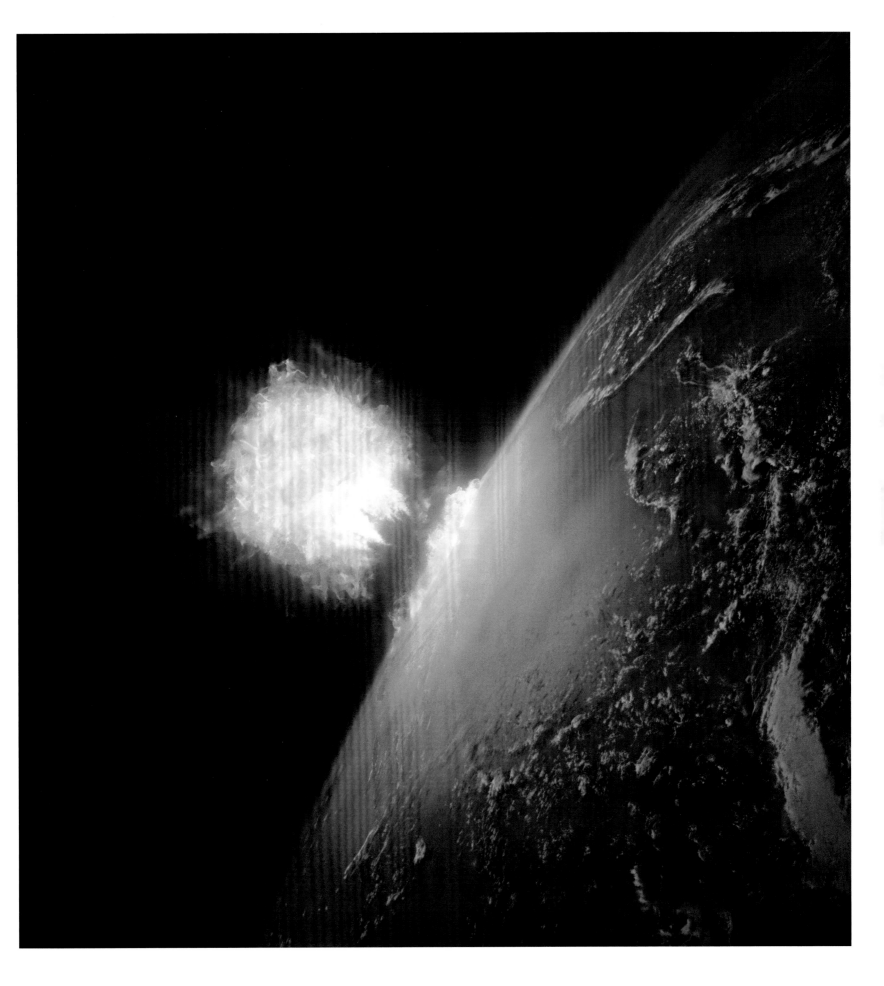

THE LONGEST WINTER

For at least six months, perhaps even for years, the Earth's surface was a deathly twilight. The asteroid had hit a patch of geology full of sulphur, smothering the planet in darkness, which then caused temperatures to plunge. The thick haze of brimstone choking the skies would have looked hauntingly beautiful, tinged with deep scarlet and crimson. As clouds formed, this sulphurous dust dissolved and fell as acid rain onto anything that was still alive below. The vegetation that had survived the blast and the subsequent tsunamis, earthquakes, wildfires, acid and plunging temperatures had to hang on with just a fraction of the light they had once received. A few dinosaurs and pterosaurs may have been able to eke out a living for a short while, but their days were numbered. It was a hopeless plight. The pterosaurs became entirely extinct during this episode, and only a fraction of the lizard and snake species would make it through the endless winter. Precious few birds represented the last surviving dinosaurs of a 150-million-year-old dynasty.

In the oceans, the lack of sunlight had shut down the photosynthesis of phytoplankton, collapsing the food webs of anything living near the surface. Ammonites — those coil-shelled cephalopods that had witnessed the entire dinosaur story — finally became extinct. Delicate corals suffered incredible losses in the shallows, with fewer than half of species

PREVIOUS PAGE: A dust bowl of ash and destruction follows a forest fire 66 million years ago. It was ignited by the searing heat of chunks of asteroid ejected from the impact site.

RIGHT: In almost complete darkness, the Earth's surface was eerily quiet, littered with the burned remains of the last giant dinosaurs.

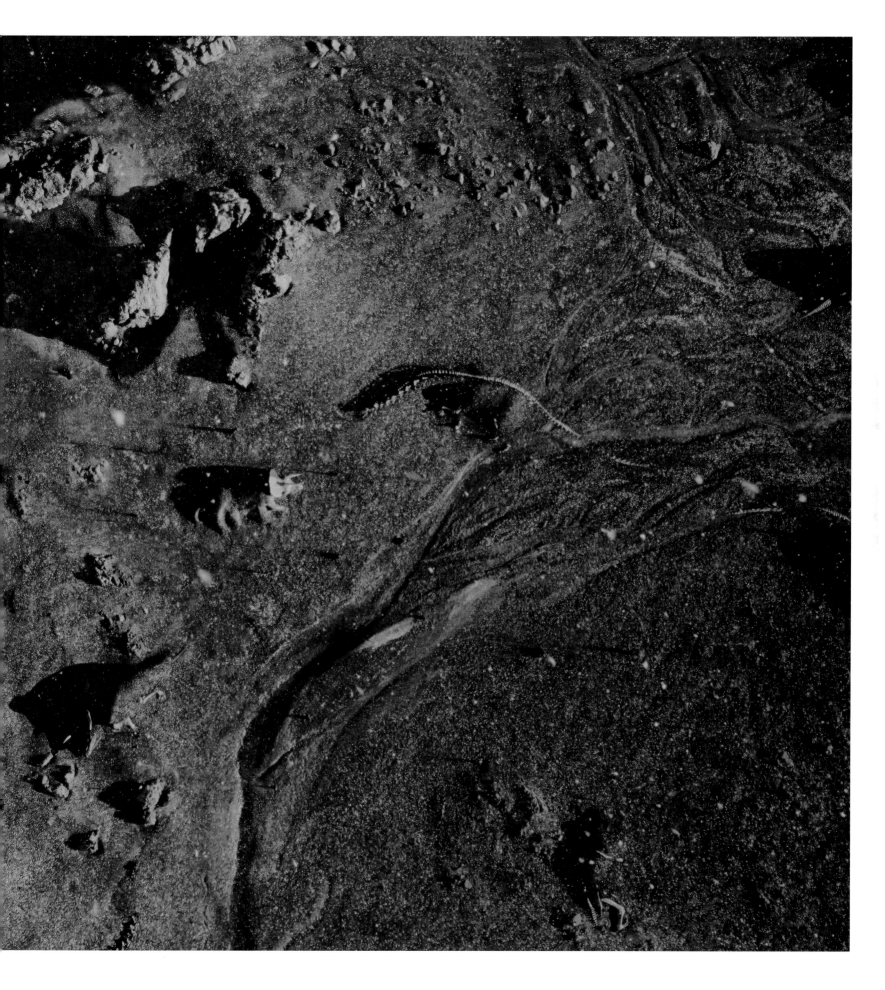

making it through and those reliant on sunlight all but destroyed. Like the dinosaurs on land, marine reptiles faced cold exposure from the nuclear winter, as well as lingering starvation. The giant mosasaurs and the last of the long-necked plesiosaurs disappeared. Deepwater creatures, less vulnerable to the hell on Earth above, may have fared better, but nothing alive was truly protected from the fallout.

On land, animals smaller than 10 kilograms had a better chance than most, perhaps because they were able to hide or needed fewer resources to survive. Streams and river systems also seem to have provided a lifeline for many animals, with amphibians, crocodiles and turtles making it through. Many of these animals can survive short spells of low temperature and are capable of fasting for long periods of time, all factors that may have improved their chances. Animals with less-specialist diets, such as scavengers and omnivores, were also more likely to survive. The post-apocalyptic menu was decidedly limited and may have consisted of just fungi and invertebrates feeding on the decay. The surviving mammals were about the size of rats, unspecialised and free to make the most of this chaos.

For around 1,000 years after the impact, ferns dominated the landscape – disaster species capable of surviving, even thriving, in the low light, and now having less competition from trees. Slowly, the dust settled and recovery could begin. It took around 200,000 years for mammal numbers to creep up, as forests recovered and habitats were rebuilt from the ground up. Angiosperms had been hit hard by the extinction, but they were in a unique position to seize power, perhaps because of their seeds. In 2012, for example, scientists in Russia discovered seeds of campion flowers buried by a squirrel almost 32,000 years ago. Despite having been frozen under ice for such a long time, some seeds germinated, instantly placing them amongst the oldest plants on Earth. This ability for seeds to survive in stasis like this could have made all the difference, a survival pod ready to activate when conditions improved. After the extinction, temperatures gradually rose and the Earth, once again, became tropical and wet, then within 6 million years dense forests consumed the land. There was no ice at the poles, no savannahs, open woodlands, shrublands or grasslands, just a tapestry of forests stretching across the continents.

It was around this time, 65 million years ago, that India, an island continent that had been merrily drifting northwards since the Cretaceous, finally began to collide with Asia. The Tethys Sea between the Indian and Asian tectonic plates began to pinch closed as the landmasses above crumpled together. The crash slowed the Indian plate by 15 centimetres a year, but even this impact would not stop the relentless push. When tectonic plates collide, most of the action

LEFT: The Redwood National Park, California, USA. Adapted to low light and cold conditions, ferns and conifers like these may have been the first plants to recover after the impact.

happens below ground, and in this case the Indian continent was being dragged below the Asian plate. Around 60 million years ago, as the conveyor belt of the Indian plate travelled deeper into the Earth, it began to melt, and light bubbles of magma started to float upwards. It is hard to imagine liquid rock melting its way up to the surface, rather like a lava lamp, but some of the most impressive volcanoes on Earth are formed in this way. While most volcanoes at the sutures of tectonic plates bubble and splatter away with a constant fizz, it is the ones that slowly make their way to the surface that are the most dangerous. While underground, the magma of these volcanoes has time to melt lots of silica, which makes it much thicker and stickier. As this viscous magma gets closer to the surface, the pressure around it decreases and the gas dissolved in the magma expands. If that gas can't escape quickly, then the mixture of gas and lava can explode outwards with incredible force. It is like opening a soda bottle containing carbonated water compared with carbonated treacle. Violent and explosive volcanism was common along the Indian–Asian crash zone at this time, and these eruptions would have been witnessed by the first animals to cross over between these newly connected landmasses. In the wake of a cataclysmic astronomical event from above, the ground itself was beginning to stir below.

OUT OF THE SHADOWS

The first 10 million years of the post-asteroid world was a time of recovery, a geological period known as the Palaeocene. The forests were fundamentally different in character than before the extinction, with angiosperms racing to fill the vacant niches. The rainforests began to look more modern, with a closed canopy and shaded forest floor. Competition for light and nutrients would become so intense that some plants even began growing, not in the ground, but on the branches of trees. The thick jungles were a wonderful playground for the evolution of surviving groups like birds and mammals. Complicated and productive habitats mean that there are plenty of places to hide, and a wealth of unique relationships to build with other species. Although they stayed small at first, mammals were quick to diversify and grow larger. The biggest mammal of the Cretaceous was about a metre long, but throughout the Palaeocene some placental mammals were reaching the size of bears. Despite looking quite wolf-like, these predators were more closely related to cows, hippos and whales. Mammals, generally, were free to evolve without the ecological oppression of the giant dinosaurs, and so in the dense forests many became fruit specialists, and began feeding without fear in broad daylight. This frenzy of diversification in the Palaeocene saw major splits in the mammal family tree, and the evolution of most of the mammal groups we know today. While egg-laying monotremes became restricted to South America and Australia, the marsupials and placentals were taking over the world.

One of the most impressive modern groups of placental mammals can trace their roots to Africa at this time, the Afrotherians. Today, the Afrotherians include an incredible array of species, and from appearances alone it is hard to believe they are related at all. Jumping shrews, manatees, aardvarks, elephants, hyraxes, tenrecs, golden moles and more all belong to this group, filling a huge number of ecological niches. All evolved from a small stock of placental mammals that must have become genetically isolated in Africa shortly after the asteroid hit. On

BELOW: Despite their rodent-like appearance, the bush hyrax of Ethiopia are afrotherians and more closely related to elephants.

the other side of the Atlantic, South America was ruled by another great dynasty of placental mammals called the xenarthrans. This ancient group includes armadillos, sloths, anteaters and impressive extinct species like car-sized relatives of armadillos and 6-metre-long ground sloths. To the north, a huge group of placentals known as the boreoeutherians were diversifying rapidly, too, having split into two groups during the Cretaceous. One group included the ancestors of hoofed mammals, carnivores like dogs and cats, many insectivores, and bats. The other gave rise to the ancestors of rodents, rabbits, and an especially important group from our perspective, the primates.

THE FIRST PRIMATES

The lush tropical rainforests of the Palaeocene were a rich feeding ground for the earliest primates, which at first resembled large rodents. By specialising on fruit, insects and young leaves high up in the canopy, they could avoid many of the predators below. However, this lifestyle presented many challenges, in particular the need to move around a lot to find food such as fruiting trees. Primates responded by evolving flexible shoulder joints for swinging, long prehensile tails, and hands with sensitive fingertips capable of grasping branches. In order to locate ripe fruit and young soft leaves, they began using their sight much more than their sense of smell. By positioning the eyes at the front of the skull, early primates could judge the distance of food without moving their heads. This ability, called 'stereoscopic vision', also allowed primates to quickly coordinate their movements through the tangle of branches. They also developed better colour vision, something that had been lost in primitive nocturnal mammals, which they didn't need while foraging in the darkness. Colour vision allowed these primates to see how ripe the fruit was, and to find the softest new leaf growth amongst a sea of green.

Primates today belong to two main groups: one contains tarsiers, lemurs and lorises; the other includes monkeys and apes. The lemur relatives could be described as more primitive, because they have smaller brains and longer snouts and tend to be more nocturnal, with poorer colour vision. The monkeys and apes tend to have bigger brains and use a wider range of habitats, rather than just living in trees.

However, the most remarkable feature of primates is their intelligence and their seemingly limitless curiosity. The ability to learn quickly and solve problems is not unique to primates, but it has been a key factor in their long-term success. It allowed them to explore innovative survival strategies, and in some cases build up complex relationships in social groups. Primates, like

RIGHT: The spectral tarsier is one of the most primitive of primates. Tarsiers are exclusively nocturnal and carnivorous, and this one on the Indonesian island of Sulawesi is returning to sleep in a fig tree after a night hunting.

capuchin monkeys today, live in groups of up to 35 animals, sharing jobs such as defence against predators and rival gangs. Capuchins are among the most intelligent monkeys, and are capable of remembering several steps to achieve a goal. Examples include using their tails as sponges to enable them to drink from deep wells, beating clams until they open, and even using tools. Brown capuchins can break palm nuts by placing them on flat boulders and hammering them open with smaller rocks. Applications of this intelligence don't just have to involve food, either. Capuchins are known to rub citrus peels and crushed millipedes onto their fur to help repel biting insects. With training, they are even capable of understanding the value of currency as a middle ground to reward, just as we humans know what money is worth.

MODERN BIRDS

Every bird living today can trace its roots back to the Cretaceous, possibly over 100 million years ago, but fossil evidence is frustratingly limited. Some of the earliest birds were the ancestors of chickens and ducks, soon followed by the group that would one day lead to ostriches, emus, tinamous, kiwis and the like. As odd as it is to imagine, waterfowl appear to have lived alongside the giant dinosaurs, and some were remarkably duck-like in appearance.

RIGHT: A capuchin monkey in Costa Rica's coastal forest negotiates with a clam. This group has learned that patiently hammering the shells eventually causes them to open.

The end-Cretaceous saw the old guard of toothed birds like *Hesperornis* and *Ichthyornis* disappear, but the ancestors of modern birds seem to have been relatively unaffected. But why? Some believe that their feathers, physiology or ability to fly helped them through, or perhaps that they were living next to rivers and lakes. However, this does not explain why animals with those same traits — such as pterosaurs and the older birds — died out. While all of these things may have played a part, the latest evidence suggests intelligence may have been crucial for modern bird survival. Preconceptions of birds as dim-witted or, dare I say 'bird-brained', are far from the truth. Birds in the crow family are known to use tools, and parrots are very capable of learning human vocabulary. Even birds we might consider stupid, like pigeons and chickens, are capable of quite complex tasks. There is also evidence that the ancestors of modern birds were evolving larger brains and a much keener sense of smell before the extinction. In the post-apocalyptic winter, flexible behaviours, curiosity, and thinking around problems would have made all the difference.

Very shortly after the mass extinction, bird diversity exploded, as they rapidly filled all kinds of empty ecological niches. New groups included the first penguins, owls and vultures, and even flightless giants like *Gastornis*. Standing at 2 metres tall, with a huge beak and head, *Gastornis* was an intimidating animal, despite being an herbivore. To the south, in what is now Colombia, lived another Palaeocene giant, the largest snake that has ever lived, *Titanoboa*. In life, it would have looked like a bulky, 13-metre-long anaconda, however, like *Gastornis,* it probably was not as fierce as its appearance might suggest. It lived in river systems surrounded by thick forests, and mainly ate lungfish and, perhaps, the occasional crocodile. Alongside *Titanoboa* there was a 3-metre-long turtle called *Carbonemys*, and, while we might think of turtles as gentle creatures, the jaws of *Carbonemys* were strong enough to eat crocodiles and mammals. Modern turtles are capable of bites with surprising force and speed, and species like the alligator-snapping turtle can easily crush bone. The size of these reptiles was likely a result of the high temperatures in its habitat, which were around 33°C on average; wonderful conditions for monstrous cold-blooded reptiles, and part of a warming trend that was happening globally.

THE EOCENE GREENHOUSE

By the end of the Palaeocene, 56 million years ago, global temperatures were on the rise, perhaps due to increased volcanic activity as the Atlantic Ocean widened. This boost in temperature of around 8°C occurred over less than 20,000 years, supercharging evaporation from the oceans. This created powerful storms that drove deep into the continental interiors, causing much of the land to become considerably wetter. At this time, ocean currents were still quite different to those we see today, but heat was being transported quickly all over the world. Even the Poles had a subtropical climate, and like previous greenhouse events, this may have prompted the release of the methane frozen on the sea floor. Methane is a very potent greenhouse gas, and had likely contributed to massive extinctions in the past. However, on this occasion the warming seems to have been a double-edged sword. While some groups of plankton suffered in the oceans, primates and other mammals were diversifying rapidly. Despite the high temperatures, life on land was apparently thriving, perhaps protected from the worst effects of the heat by the verdant, moist forests.

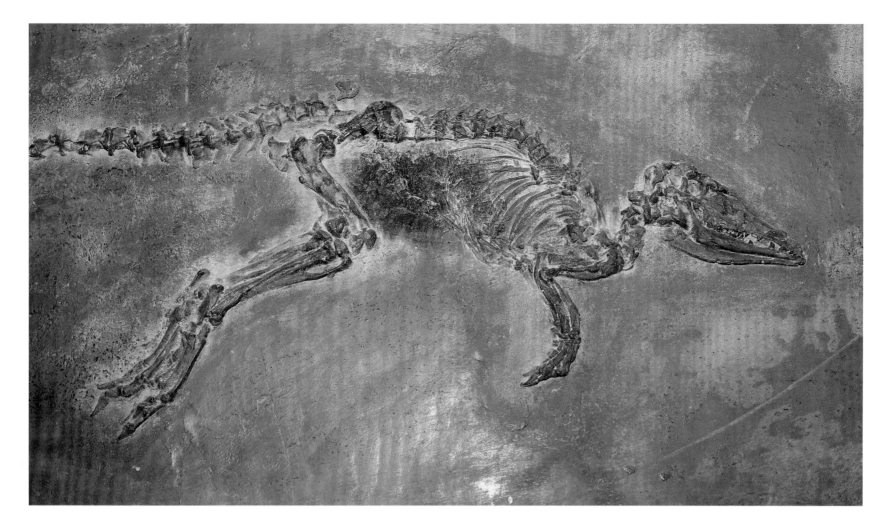

The beginning of the next geological period, the Eocene, remained warm and wet and saw the first appearance of hoofed mammals, or ungulates. These early ungulates were generally small, adapted for low-lying browsing in the dense forests. Among them were even-toed ungulates, rather like early horse and rhino relatives, no larger than around 1.5 metres. Odd-toed ungulates were incredibly small at this time as well, around the size of rabbits, which is quite incredible considering that today this line of mammals includes the blue whale. These early ungulates were soon joined by predators, such as the creodonts, a now-extinct group of carnivores that resemble dogs and cats in appearance and size. The small size of all of these animals may have been in response to the high temperatures, which rivalled the sweltering highs of the Cretaceous greenhouse, and even the intense global warming of the Permian–Triassic extinction.

In one particularly spectacular fossil site in Messel, Germany, a snapshot of the Eocene forests is captured in splendid detail. The rocks that can be found there today are so delicate that the fossils have to be covered with resin before they are painstakingly revealed with fine tools. Over 1,000 species have been recovered from the pit, which was narrowly saved from

ABOVE: *Leptictidium* was a small hopping mammal found in the forests of Europe during the Eocene period.

becoming a landfill in the 1990s, and is now a UNESCO World Heritage Site. During the Eocene, around 47 million years ago, the Messel Pit was a deep lake surrounded by thick forests. At the bottom of the lake, the water was stagnant and perfect for preserving any animals that drifted down into its depths. Crocodiles, fishes, salamanders and frogs all fell victim. However, as well as the aquatic animals that you might expect to find in a lake, there are also an unusually high number of land-living creatures preserved there. These include hedgehog and aardvark relatives, pangolins, marsupials, primates, rodents, insects, spiders, pygmy horses and a large array of birds, including that large bird *Gastornis*. The Messel site also preserves fossils of a charming little hopping mammal called *Leptictidium*, which has no living descendants. The preservation is so good that the outlines and markings of its fur can be seen, as well as the leaves, insects and small vertebrates in its stomach. As beautiful and informative as these fossils are, the question remains, why are so many preserved in this lake? Indeed, Messel had a sinister secret. It is now believed that volcanic activity deep under the lake occasionally caused the release of carbon dioxide into the air. This gas would have asphyxiated anything in its path, as the dense invisible cloud rolled through the trees. It would have been a haunting scene as the forest grew eerily silent and the bodies of all manner of fauna fell from the air and treetops. Among the victims discovered in this graveyard lake are freshwater turtles that died while mating; in fact, nine pairs of copulating turtles have been fossilised, clearly overcome by the toxic gases.

RECLAIMING THE SKIES

Some of Messel's most beautiful treasures are the delicately fossilised bats, which even preserve the wing membranes between their fingers. Bats appeared in the fossil record a little earlier than Messel, in North American lake deposits dating to around 53 million years ago. Even then, they already had wings, slept upside down and resembled modern bats, so they must have evolved much earlier than we have the fossils to prove. These Eocene bats were probably also capable of sonar echolocation, producing high-pitched sounds and listening for their echoes to build up a picture of their surroundings. This ability may have started evolving back in the mammals' earliest days (or rather nights), to listen for insects hidden in the darkness. Echolocation allowed bats to fly at night and navigate in complex environments, opening up huge opportunities for hunting. Bats, today, are incredibly diverse, living everywhere except the poles, and accounting for about a fifth of modern mammal species. Most eat insects, but many others are fruit or nectar specialists, which are important pollinators and seed dispersers. Some are capable of catching fish, frogs and other small vertebrates on the wing, and notoriously some even feed on the blood of other animals.

Without fossils earlier than the Eocene, it is difficult to know quite how bats took to the air, but their reason for doing so is clear. With the flying reptiles extinct, and bird numbers at a fraction of what they were, the post-asteroid skies were there for the taking. The race to claim the skies was essentially reset at the end of the Cretaceous, and bats and birds have been vying for dominance ever since. Despite the variety of species, most bats alive today are nocturnal, and this is very probably because of predation by birds of prey. In daylight and in open environments, birds of prey are capable of building up speed and plucking bats out of the air with their huge talons. Bats are, in fact, much more acrobatic fliers in the technical sense, because they can use any one of their long fingers to alter the shape of their wing membrane while in flight. Birds, on the other hand, have a simpler wing, which generally favours speed over manoeuvrability. Many birds of prey, or 'raptors', can trace their roots to the Eocene forests or earlier, including the ancestors of hawks, eagles, vultures and kites. Falcons also evolved in the early Eocene but, despite having a similar appearance to hawks, are actually more closely related to parrots. Anyone who has met a captive parrot can attest to their falcon-like character, and might not find this surprising.

Candidates for the first birds of prey to evolve after the asteroid were the owls, which date back to around 60 million years ago. These early forms probably hunted during the day and only became nocturnal specialists relatively recently. Owls have truly embraced the night, with a suite of adaptations for flying and detecting prey in near darkness. They fly relatively slowly, but the unique structure of their wings reduces the sound they make, and even alters the pitch of any noise produced to below the hearing range of their prey. Unlike bats, which use echolocation, owls are highly visual and have powerful eyesight, even in darkness. Light sensitivity is maximised in a similar way to many nocturnal mammals, using a reflective membrane behind the retina, called the tapetum lucidum. The eyes themselves are so big that they have evolved to be tube-like in order to fit into the small skull, and because of this tight squeeze owls cannot move their eyes around like we can. To compensate for this, they have twice as many neck bones as humans, which allows them to turn their heads up to 270 degrees. Their hearing is also exceptional and may be their primary sense in very low light conditions. Some have ears placed asymmetrically on the skull, to determine the height and direction of prey by detecting tiny differences in sound arrival time. The owl pinpoints the target by turning its head, until

RIGHT: Honduran white bats of Central America construct tents for shelter. They cut the leaf veins with their teeth so the leaf can be more easily bent over.

NEXT PAGE: A martial eagle soars through a cloud of straw-coloured fruit bats in Kasanka National Park, Zambia. The site serves as a seasonal roost for 5–10 million bats each year.

soundwaves reach both of the ears at the same time. Many owls also have a disc of facial feathers around the face, which acts like a satellite dish, focusing sound into the ears. Night or day, the birds and mammals had truly reclaimed the skies and filled the void left by pterosaurs and the extinct orders of ancient birds.

THE DESCENT OF WHALES

Throughout the Eocene, tectonic plates continued to push and pull continents into their modern configuration, but there was a lot still to come. To the north, Europe, North America and Asia were connected, and dominated by placental mammals. To the south, Antarctica, Australia and South America were also connected, but they were ruled by marsupials. Africa may have been isolated, with a narrowing sea called the Tethys at its northern edge. These were highly productive waters, full of plankton and other microscopic creatures which made their shells out of calcium carbonate. One of these was a single-celled organism called Nummulites, which had a round shell that could reach 10 centimetres or more, a giant of its kind. Its name derives from the Latin for 'little coin', and the ancient Egyptians used to use these fossils as coins. As the carbonate shells of Nummulites piled up on the Tethys sea floor 50 million

RIGHT: A great grey owl flies silently through a pine forest in Finland. Its excellent hearing allows it to locate small rodents, even beneath thick snow.

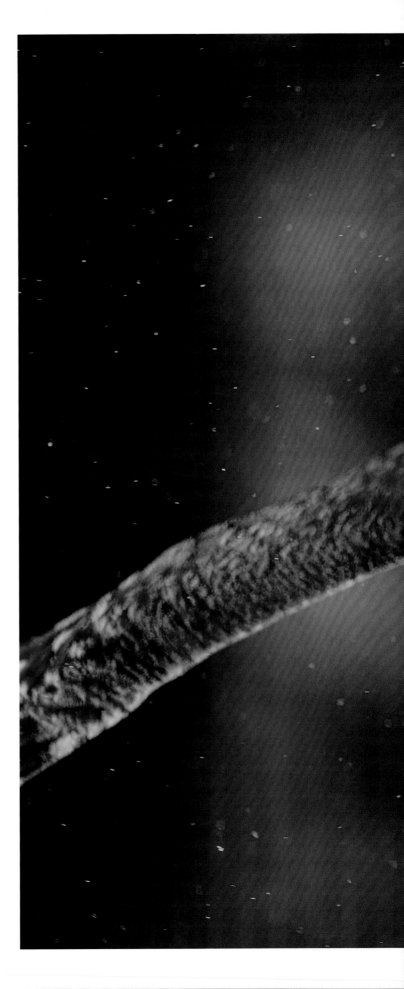

years ago, they became limestone, which was later used by the ancient Egyptians to build the pyramids.

It was on the shorelines of the Tethys Sea that one group of mammals took a giant evolutionary leap into the water. Thanks to genetic studies, we now know that whales' closest living relatives are hippos, and both belong to the even-toed ungulates, a hoofed order of mammals that today also includes cows, deer, pigs, camels and giraffes. The fossil evidence has been improving greatly over the last few decades, and we now have a good idea of how the transitional ungulate-to-whale forms looked and behaved. Thanks to a close relative of the earliest whales, *Indohyus*, we know they evolved in the early Eocene from an omnivorous mammal, no bigger than a raccoon. Like whales today, the ungulate *Indohyus* had a special thickened ear bone, which helped it hear underwater, and dense bone in its legs, which helped it walk across the sea floor or river bottom. Eventually creatures similar to *Indohyus* would lead to true whales, the cetaceans, and lose the slender legs and hooves. However, life in water presented many challenges – not least the huge effort required to move through a fluid that is 800 times denser than air.

Modern otters, platypus, beavers and the like benefit from the safety of land, but they also have enough agility and sensory power to swim and feed underwater. Today, seals and their relatives demonstrate just how skilfully mammals can manoeuvre underwater, but even they must return to the shore to rest, mate and give birth. Around 50 million years ago, after small ungulate forms like *Indohyus* came the wolf-like *Pakicetus*, regarded as the first cetacean or true whale. It had teeth adapted for eating fish and, probably, patrolled rivers and lakes near the sea. *Pakicetus* and its relatives belonged to the most ancient group of whales, the archaeocetes, which became more aquatic throughout the Eocene.

By 47 million years ago, archaeocetes like *Maiacetus* were semi-aquatic, and ecologically tied to the oceans. It had a streamlined head and body, and cone-like teeth similar to modern killer whales. The name itself means 'mother whale', because of one extraordinary fossil specimen discovered with an unborn foetus within an adult's ribcage, about where the womb would have been. It faces towards the tail end of the mother, and if this was its natural orientation, then it could be a clue that *Maiacetus* gave birth on land. Modern whales and dolphins give birth to their young tail-first, which leaves the umbilical cord intact for longer during difficult births and reduces the risk of the baby drowning. This is seen in ichthyosaurs as well, which despite being ancient reptiles clearly faced the same problems of giving birth underwater.

Maiacetus was about the size of a person, and had quite short limbs which would not have been much use on land. In the water, *Maiacetus* would have undulated its whole body to swim

LEFT: Recent DNA analysis has revealed that hippopotamus are the closest living relatives of whales and dolphins.

NEXT PAGE: A group of *Maiacetus*, early whale ancestors, sunbathe on the shores of the Tethys Sea 47 million years ago.

and paddled with its legs. As archaeocetes evolved, they began using their front limbs and tails much more to generate thrust. The hind limbs became smaller and they evolved wider tails. The most impressive of these archaeocetes was undoubtedly *Basilosaurus*, a huge predator of the late Eocene that measured up to 20 metres long. It had almost completely lost its hindlimbs, and had well-developed flippers and a tail fluke, making it among the first whale groups to become fully aquatic. *Basilosaurus* was likely the top marine predator of the Late Eocene Tethys, and may even have fed on other related whales.

Whales had truly transformed their body plan throughout the Eocene, from animals the size of racoons to the giant *Basilosaurus* in just 7 million years. However, whales were not the only marine predators that had been getting scarily huge. The group of giant Cretaceous sharks, which may have made a meal of dinosaurs, had survived the asteroid. Their descendant was *Otodus*, a distant relative of the great white shark but twice the length. Throughout its evolution, *Otodus* had been growing, and its teeth were changing to become much more robust, serrated and triangular. This shark's diet was clearly changing from small fish to very big mammals. As whales grew, so did *Otodus*, which would remain their main predator until the bitter end. *Otodus* finally became extinct around 4 million years ago, and

RIGHT: The apex predator *Otodus obliquus*. This huge lamnid shark would one day give rise to the infamous megalodon.

arguably saved the best till last. *Otodus megalodon* was the last of its kind, but may have grown over 15 metres in length, with teeth the size of hands and a bite force more powerful than that of *Tyrannosaurus rex*.

BIRTH OF THE HIMALAYAS

We must return briefly to those early days of the Eocene, when hot forests were blossoming with diversity and the tectonic collision of India with Asia had been thundering on. At first, there had been fierce volcanism, as the Indian plate melted below the Asian plate and bubbled up to the surface, but around 55 million years ago this began to stop. Like a slow-motion car crash, the zone between the two landmasses was crumpling, mashing the two enormous slabs together. Some of the rocks buckled, squeezed into giant folds like a wrinkled rug pushed across a polished floor. At other times the rocks simply broke and pushed past each other along giant faults. With pressure from every side, there was nowhere for the rock to go but up. Mountains began to build, pushing upwards by 160 metres every million years. These were the Himalayas. Eventually, the mass of scrambled rock piled so high that even bubbles of magma from the melting Indian plate could no longer reach the surface. The violent volcanism they had caused subsided. As the Himalayas grew taller, rock that had previously been at sea level was carried skywards. Of course, the most famous and tallest mountain of the Himalayas is Mount Everest, at over 8,848 metres high. At the peak of Everest there is a sedimentary rock called limestone, laid down in a shallow sea possibly over 500 million years ago. Rocks collected just 6 metres below the summit contain ancient marine animals like trilobites, sea lilies and even fossilised faecal pellets. It was quite the journey for those remains; from the bottom of a Cambrian sea to the highest point on Earth.

As well as marine sediments, a lot of igneous rocks – solidified magma or lava – were also brought to the surface by the mountain building. As the Himalayas grew, so did the amount of rock exposed to the elements. The weathering acted like a chemical sponge for the greenhouse gas carbon dioxide, helping to reduce global temperatures. At the beginning of the Eocene, the climate had been a hot greenhouse, but the formation of the Himalayas may have turned things around. On its own, this would have been a gradual process, but there were other forces at work that caused more dramatic changes. These included the regular ups and downs associated with Earth's orbit and shifting wind and ocean currents. However, one of the most potent occurred in the middle Eocene around 49 million years ago, when carbon dioxide plummeted by 80 per cent in just 800,000 years. The culprit, surprisingly, was a 2-centimetre freshwater

RIGHT: The Himalayan mountains Cholatse and Taboche tower above the Gokyo Lakes, the highest bodies of freshwater on Earth, in the Sagarmatha National Park, Nepal.

fern called *Azolla*. At the time, the Arctic sea was cut off from the oceans, and so when water evaporated, the saltwater below became concentrated, making it more dense. When rain filled the rivers and replenish this ocean, the fresh water sat on top, like a layered cocktail. This thin layer of freshwater, rich in nutrients washed off the land, was enough to provide a home for the floating *Azolla* ferns. In the summer months they would grow rapidly, doubling in mass every few days, and covering the ocean in a thick green mat of plant matter. As they photosynthesised during the long Arctic summer days, these mats were drawing down 1.5 kilograms of carbon dioxide per square metre each year. The crucial part of this story is what happened to those mats after the summer ended. As the nights grew longer and colder, and the *Azolla* began to die off, the mats sank. In normal circumstances, the decaying plant material would be consumed by aquatic organisms, and recycled back into the environment. However, the dense saltwater of the Arctic sea was stagnant at the bottom, so the *Azolla* and all its carbon could pile up, undisturbed. Gradually sediment would cover them, and lock that carbon away.

ABOVE: The aquatic fern *Azolla* can double its mass in less than two days, locking away atmospheric carbon.

THE ICE RETURNS

Throughout the rest of the Eocene period temperatures continued to drop, and the forest blanket of warmth and humidity finally began to lift. Many parts of the world started to have wet and dry seasons, and the North and South Poles became cooler and more arid. Those animals that could migrate away from the Poles did, forced into the remaining rainforests at the equator. By around 40 million years ago, many of the older groups that could not adapt had gone extinct, including an ancient offshoot of the rodents and a huge number of primates living in Europe and North America.

The Himalayas continued to weather and draw carbon dioxide out of the atmosphere, but, at the time, this was not the only thing affecting global temperatures. Tectonic forces were also changing the configuration of continents, and the way ocean currents transported heat around the world. For a long time Antarctica had been connected to South America and

ABOVE: Large mats of *Azolla* grow in Oaks Bottom Wildlife Refuge, Oregon, USA. Remarkably, these tiny plants may have caused global cooling in the middle Eocene.

Australia; however, Australia had been moving steadily northwards away from Antarctica. By around 35 million years ago, a seaway had opened up between the two continents, known as the Tasmanian Passage. At around the same time another seaway – the Drake Passage – opened up between South America and Antarctica. Although these openings were narrow, they fundamentally altered the ocean currents in the Southern Hemisphere, and beyond. With the Drake and Tasmanian Passages open, water could move unimpeded around Antarctica. The Antarctic Circumpolar Current was born, the most powerful ocean current on the planet. It travels at around 180 million cubic metres of water per second, which is 100 times more flow than all the rivers on Earth combined. The Drake Passage is a pinch point for the current, at just 800 kilometres wide, and huge eddies form as water passes through it from west to east.

The effect of the circumpolar current was to isolate Antarctica, encircling it and preventing heat from entering the region, or cold from escaping it. At this time, carbon dioxide was dropping to a threshold level where ice could form, and it is possible that Earth's orbit helped nudge temperatures down as well. A more dramatic cause for the drop in temperature has also been suggested – asteroid strikes. Around the time of the cooling, two asteroids hit Earth within quite a short timeframe. The first hit Popigai in Siberia 35.7 million years ago, creating an impact crater 100 kilometres wide. It smashed to Earth with such force that carbon in the ground was instantly turned into diamonds. A second smaller asteroid struck Chesapeake Bay, USA, around 35.5 million years ago, with a 40-kilometre-wide crater. These asteroids would have hit at 18 kilometres per second, hurling ash into the atmosphere and temporarily blocking out the Sun. If the impacts were big enough, they could have drastically reduced global temperatures. As it stands, we simply do not know what ultimately caused Antarctica to freeze, but once this process began, it would have been hard to stop. As snowfalls and freezing lakes turned the land white, Antarctica began to reflect more light from the Sun. Snow reflects around 80–90 per cent of the Sun's light and heat, whereas water reflects about 6 per cent and absorbs 94 per cent of that radiation. This is called the albedo effect, and it would have greatly accelerated the freezing of Antarctica and the development of ice sheets and glaciers 34 million years ago. As freshwater froze and accumulated at the South Pole, it was effectively locked away from the hydrological systems of the rest of the planet. The shock of this cooling climate and drastic aridification of the Earth would have huge impacts on life. In the 30 million years since the apocalyptic asteroid strike, Earth had been a tropical greenhouse, fostering diversity that exceeded anything that had come before. The icehouse climate that lay ahead would change the world and test the survival of every living thing.

RIGHT:
The Antarctic Circumpolar Current is the largest ocean current on Earth. Its formation around 35 million years ago helped ice sheets form at the South Pole.

THE AGE OF ICE AND FIRE

The beginning of the Eocene had seen global temperatures spiking to around 25°C, but by the end of the period this had plunged by 8°C. Earth was entering an icehouse climate, and with lower temperatures came a great drying of the land. About 36 million cubic kilometres of Earth's freshwater was gradually locked away as ice as the temperatures plummeted further. Swathes of Asia and North America became more arid, and the plants that were adapted to warmer and wetter conditions began to vanish. Tree-dwellers, such as primates and browsing animals, would be particularly badly affected as their habitats and food sources dissolved away. The next geological period, the Oligocene, would be a time of transition, as modern families emerged and archaic forms disappeared.

MODERN WHALES

The isolation of Antarctica had not only dried the land but also affected ocean currents all over the world. As these great conveyor belts of water shifted, they hit land masses in different ways. Some began to bear the brunt of underwater storms, which swept nutrients up to the sunlit surface, fuelling huge blooms of plankton. In other areas, waters that had previously teemed with life grew silent and barren as their upwelling currents petered out. These new productive areas brought feeding opportunities that whales were in a unique position to exploit. By the end of the Eocene, whales had split into two main groups: the toothed whales and the baleen whales. Toothed whales include animals like dolphins, orca and sperm whales, which, as the name suggests, have teeth. They are also capable of echolocation, making them efficient hunters even in murky or deep, dark waters. The baleen whales are so called because of the rows of hairlike material in their mouths, which is used to filter food out of the water. Baleen whales probably began as relatively small animals with teeth that used suction feeding to forage on the sea floor. In time, the suction behaviour was combined with the baleen filters as a way to consume larger quantities of smaller plankton. By the end of the Oligocene, baleen whales were filter feeding almost exclusively, and the teeth were lost entirely.

PREVIOUS PAGE: Cave lions traverse the icy tundra of the mammoth steppe 20,000 years ago.

RIGHT: A conifer plantation alongside native deciduous trees in Thetford Forest, United Kingdom.

THE FORESTS FRAGMENT

As great tracts of land became drier, plants with broad fleshy leaves gave way to scrub. Forests thinned and the canopy began to break up, and in response many herbivores became smaller, to enable them to survive on less food, and being large became an increasingly precarious strategy.

Among those struggling to adapt were the brontotheres, a group of mammals related to horses, rhinos and tapirs. These were some of the largest herbivores on Earth at the time, and species like *Megacerops* of North America could reach 2.5 metres tall, and 3.5 tonnes. Their teeth were adapted for chewing soft leaves, and so to find food they had to travel further and forage wider areas. For a while *Megacerops'* large size would have been a blessing, allowing them to use stored fat to migrate and follow the rains. However, in time even these mighty animals would become rarer and would later disappear entirely.

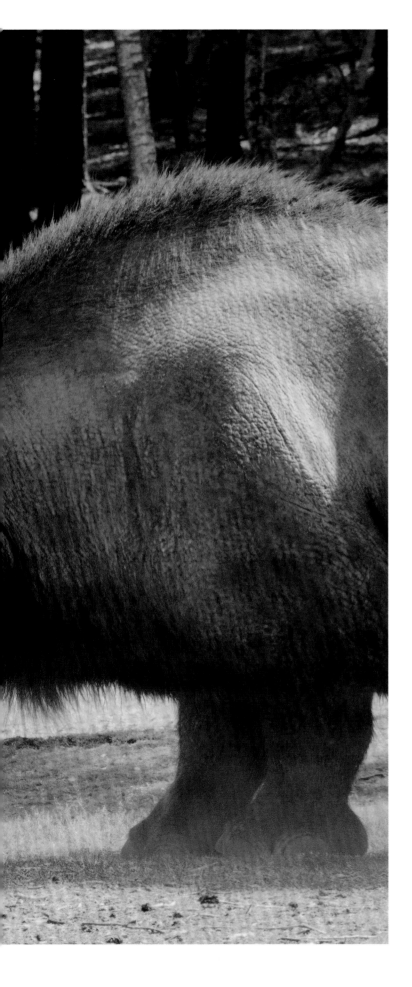

Descending temperatures were also causing sea levels to drop by over 100 metres, and this was creating its own chaos. The Turgai Sea had separated European and Asian animal populations since the dinosaurs, but as sea levels fell this natural barrier drained away. Suddenly, there was competition between mammals from both sides, and most of the European mammals did not fare well. Ecological restructuring like this was happening all over the world, as drier and cooler scrub habitats came to dominate.

In southern Asia, pockets of lush forest and lowland provided a lifeline for distant relatives of *Megacerops*. The 5-metre-tall *Paraceratherium* weighed about 20 tonnes, making it one of the largest land animals that has ever lived. This is approaching the upper limit of what is physically possible on land, but it was clearly a successful strategy for these animals. They resembled a cross between a giraffe and a tapir, with a long neck, robust body and four pillar-like legs. In many ways, they were the sauropod equivalents in this new mammal age, able to reach soft new growth high up in the trees.

Around the Tethys Sea there were other patches of tropical forests acting as refuges for tree-dwellers like primates. In the Fayum District of Egypt, primates were diversifying rapidly, alongside familiar African fauna we would recognise today. One of these,

LEFT: Two male *Megacerops* face off. They were brontotheres, large and powerful herbivores, but they could not survive the global changes in climate at the end of the Eocene.

Aegyptopithecus, was larger and slower than the other primates, with a heavy build and strong jaws. *Aegyptopithecus* was among the first of a new breed of primates called the Old World Monkeys, a group which today includes apes, baboons, macaques and the like. Rainforests of South America were providing similar refuges for a splinter group of primates, the New World Monkeys. This group includes spider monkeys, howlers and capuchins, as well as smaller species like tamarins and marmosets. New World Monkeys had split from African stock 40 million years ago, and may have crossed the Atlantic before it got too wide. By the Oligocene, they were evolving larger brains, and filling many of the same niches as their African cousins. However, unlike Old World Monkeys, they never left the trees and came to rely on their prehensile tails as a sort of fifth limb to help them move around quickly.

The Oligocene also saw the arrival of the first pigs, which evolved in Asia and quickly spread to North America. At the time, animals could move freely from Eurasia to North America, because there was a land bridge they could cross called Beringia. Since the Cretaceous, the bridge had opened and closed depending on sea level changes, allowing periodic migrations of animal populations from either side. Among the travellers were early ungulate ancestors of horses, which evolved in Europe during the Eocene, before spreading all over the northern hemisphere. The earliest ancestors were about the size of a dog, and they walked on four toes at the front and three toes at the back. By the Oligocene they were larger and had lost one of their front toes, putting more of their weight on a large central toe – it is the equivalent of us walking around on our middle fingers. Finally, around 10 million years ago, the one-toed horses spread to South America as well as back across the Beringia land bridge to Eurasia, then on to Africa. As their feet transformed, so did their teeth. The earliest Eocene forms had low cusped teeth for grinding up soft broad leaves, but as the forests fragmented these became longer and grew continuously to deal with the tougher vegetation. One fast-growing angiosperm, grass, was particularly difficult to digest, and unfortunately for the herbivores, it was steadily spreading all over the world.

MARCH OF THE GRASSLANDS

The grass family of flowering plants evolved in the Cretaceous, but for a long time it only made up a tiny percentage of the total flora. Cooling since the Eocene was creating the perfect conditions for grasses, which had a few advantages over other plants. Many reproduce quickly, either by producing off-shoots or masses of seed spread by the wind. They also grow upwards from the base, meaning they can survive having their top growth eaten by grazing animals, and even burning by wildfires. Grasses actually benefit from fire, which serves to remove dead and old growth, as well as other plants that might compete with it. Following wildfires, grass grows remarkably quickly, providing a bounty of new growth for herbivores. While this may seem like a disadvantage, grazers are in fact powerful allies of grasses because they help to weed out other less-vigorous plants. By the start of the Miocene period, around 23 million years ago, grasslands were becoming extensive all over the world. Grasses could adapt to the poor soils and prolonged droughts typical of savannas. Today, these habitats are

BELOW: In order to feed on abrasive grass leaves, plains zebras have high-crowned teeth and large fermenting chambers in their gut.

witness to the largest animal herds on Earth, supported by the conveyor belt of green leaves emerging from the ground.

Herbivores face a number of challenges when eating grass, not least the abrasive silica crystals embedded in its tissues. As we have seen, horses came to deal with this issue by evolving larger teeth that grow throughout their lifetime, so there is more crown material available to wear down. Horses and their relatives also have large fermenting vats towards the end of their digestive tracts, which house bacteria capable of breaking down cellulose. Other ungulates, such as antelope, cows and their relatives, chew the grass a little before swallowing it into fermentation vats in their stomachs. After it has been partially digested and softened up by the bacteria, they can then regurgitate the food back up into their mouths at will. This regurgitation, or 'chewing the cud', can happen whenever the animal feels safe, meaning there is less time spent with their heads down when they are vulnerable to predation.

Since the grasslands emerged, predators have evolved in tandem with their prey, resulting in a number of extraordinary adaptations – for example, in Africa wildebeest synchronise to give birth *en masse*. These are not easy meals for predators, though, as large herds mean many eyes are always watching out for them. Identifying weak or young targets within a group of hundreds or thousands is no small task, and even then it is easy to lose them in the crowd. Wildebeest calves are able to walk within minutes of being born, and in this brief window of vulnerability, the parents can be incredibly defensive. In response, predators like lions have evolved to work cooperatively, which is unusual for a big cat, while other species, such as cheetah, have become fine-tuned for blistering speed. This kind of arms race is not unique to Africa. Grasslands today cover 40 per cent of the planet's land surface, where both predators and prey have been forced to adapt to the open, treeless expanses.

MIOCENE LIFE

The Miocene saw a wave of modern mammal families appear, including the first giraffes, antelopes, buffalo, hyaenas, modern-looking bears and wolves. In North America, a small, semi-aquatic mammal called *Puijila* was about to take the plunge. Although superficially otter-like in appearance, *Puijila* was more related to bears, and swam in a different way. It had large webbed hands, which it used to propel itself through the water, relying much less on its tail for manoeuvres. This was, in fact, the earliest seal, the last of many land-living groups to re-enter the ocean. Throughout the Miocene, early seals spread across the northern hemisphere,

and their limbs became more flipper-like. Unlike dolphins, which are built for speed, seals need to carefully control their movement, and the flippers were perfect for hunting slower-moving animals in the shallows. Indeed, the open ocean was no place for a small seal in the Miocene; deeper waters belonged to megalodon sharks and predatory toothed whales, some of which had biting teeth in excess of 36 centimetres long, the largest in history.

Like the rest of the world, South America was undergoing dramatic climate changes, but the cooling and drying were especially severe there. A combination of changing oceanic currents and mountain building on the west coast deprived the continent of rainfall. As the Andes grew, many of the forests of South America began to fragment, replaced by pampas grassland. This mountain building caused such upheaval that it even reversed the direction of the Amazon River – since the Cretaceous, the Amazon had drained to the west to the Pacific Ocean, but today it runs east, draining out into the Atlantic.

South America was still isolated at this time, and it was home to a menagerie of very odd animals. Among them were giant armadillos, ground sloths and strange ungulates (hoofed mammals) called litopterns, which are like nothing else on Earth. Litopterns resembled llamas but genetic studies have shown that they were more related to horses and rhinos. Many of these

ABOVE: A grey seal swims in the cold Atlantic waters of the Blasket Islands in County Kerry, Ireland.

ABOVE: A herd of *Theosodon* browse in a clearing. These strange ungulates were unique to South America, living around 15 million years ago.

litopterns began losing toes to enable them to run faster on firm ground, and some even got down to a single toe, mirroring the evolution of horses to the north.

Predators in these environments included large carnivores belonging to a particularly ancient line of marsupials. Despite being completely unrelated, some of these marsupials resembled wolves, bears, wolverines and even sabre-tooth cats. However, the most impressive predators of Miocene South America were the aptly named terror birds. Species such as *Phorusrhacos*, stood at over 2 metres tall, with a head over half-a-metre long. This bird may have been able to run at 50 km/h, about the same as an ostrich. It was armed with a robust beak with a sharp downwards-pointing tip and, based on its living relatives, the seriemas, this was probably used as a weapon. Experts believe it hunted by running alongside prey, swinging its head downwards to strike and stabbing it with its beak, like a pickaxe. As if that was not enough, it also had

strong legs and clawed feet that could deliver a powerful and painful kick, much like modern cassowaries and secretary birds.

While tectonic forces brought aridity and hardship to South America, they brought a wealth of productivity to the other side of the Pacific. At the start of the Miocene, 23 million years ago, Australia began to collide with Sulawesi, closing a deep ocean which had been separating the two landmasses. This pushed up mountains, islands and even the sea floor. In a relatively short time, the region had become the world's largest shallow marine habitat, containing about 85 per cent of all living coral reefs. When the oceans first cooled, at the end of the Eocene, reef fish had suffered a prolonged mass extinction, but this new habitat fostered an explosion in their diversity, as tight links between fishes and corals developed. The Indo-Pacific was becoming the richest marine biodiversity hotspot on Earth, claiming the title from the Tethys, which had all but vanished.

ABOVE: The Miocene terror bird *Phorusrachos* was a formidable presence in the forests and grasslands of South America.

This productivity in the seas may even have sparked the evolution of fish-eating birds, such as boobies and gannets, which appeared at this time. Seabirds are important biological pumps, carrying nutrients from sea to land, and their populations are truly staggering, with an estimated 1,200 million living in the world today. Gannets, in particular, are remarkably adept at catching fish, piercing down into the water from heights of up to 30 metres, at speeds of up to 100 km/h. They even have air sacs under the skin which act like bubble-wrap, cushioning their impact with the water. Among birds, the unparalleled masters of the water are the penguins, which spread across the southern hemisphere during the Miocene. Their streamlined bodies make them the world's fastest diving birds, with some reaching over 35 km/h. Their strong wings fly underwater with ease, and species like emperor penguins can dive to over 560 metres depth, and hold their breath for over 20 minutes. This is all the more incredible when you consider the freezing temperatures and the constant threat of attack from leopard seals and orcas.

Some mammals on land were also dealing with extreme cold, and they were about as far away from the sea as it is possible to get. Genetic evidence suggests big cats originated in central northern Asia about 11 million years ago, before diversifying and spreading to Africa and North America. Fossil evidence has been found of a six-million-year-old big cat living in the Himalayas that closely resembled the modern snow leopard. Like penguins, snow leopards have made their bodies compact and well-insulated to deal with the cold, and it is likely this extinct species did the same. Modern snow leopards are elusive and vanishingly rare, spread thinly across 12 countries. Incredibly they can take down prey three times their size, and effortlessly negotiate the rocky and steep-sided valleys they call home.

OUR EARLIEST ANCESTORS

In southern Europe, advanced Old World Monkeys were spending more time on the ground and were beginning to resemble some of the lesser apes, such as gibbons. These gave rise to the hominids, or 'great apes', which today include orangutans, gorillas, chimps and humans. All were large-bodied with males tending to be bigger than females. They also lacked the special fatty pads of the rump, used to sit on thin branches. By the middle Miocene, around 15 million years ago, the orangutan line had branched off from African apes like gorillas, and by eight million years ago human ancestors had split from our nearest relatives the chimpanzees and bonobos.

The details of how the human line emerged is the subject of hearty scientific discussion, to put it mildly. However, we do know that in the Pliocene geological period, around four million years ago, our ancestors, the australopithecines, had evolved. The earliest of these was *Australopithecus*, which had strong ape-like jaws but was also able to walk upright, distinguishing it from earlier apes. Walking on two feet allowed *Australopithecus* and later species to carry things like food and juveniles, and stand taller to see over longer distances. They were still adapted to climb trees quite well, though it should be noted that one particularly famous specimen, called 'Lucy', may have died after falling out of a tree. *Australopithecus* had brains around three times smaller than ours today but, nonetheless, larger than those of their predecessors. There is also evidence that australopithecines were using stone tools and leaving scratches on the bones of animal carcasses they were butchering. Even at these early stages, they appear to have been quite skilled at sharpening and shaping materials like flint, in a process called knapping. This increased intelligence was driven by more complex interactions and social structures, and would be pivotal in the next stages of our evolution.

ABOVE: A chacma baboon climbs a tree in Mana Pools National Park, Zimbabwe.

THE AMERICAS UNITE

ABOVE: Having made it across the Panama land bridge to North America, two male *Titanis* display aggressively in a territorial dispute, firmly in the sights of a pair of *Smilodon gracilis*.

NEXT PAGE: A *Smilodon populator* puzzles over how to tackle a herd of *Doedicurus*, huge extinct relatives of modern armadillos.

For 100 million years, North and South America had been separated by a seaway that linked the Atlantic and Pacific Oceans. All that was to change as tectonic forces drove the two landmasses closer together, causing the sea floor to rise and volcanoes to erupt. As water gave way to land, the Isthmus of Panama was born, bridging the two worlds for the first time since the dinosaurs. By around three million years ago the unique faunas of north and south could finally disperse, an event so significant that biologists termed it the 'Great American Interchange'. In South America, the strange marsupials, ungulates, sloths and terror birds had been evolving in exquisite isolation, joined by the occasional rodent and primate that had managed to raft over from Africa. The newly connected continents became a battleground, and there were very clear winners. With the exception of armadillos and opossums, very few southerners prospered in the north, fewer still beyond the equatorial rainforests.

The migrations south were much more successful, and were akin to a hostile takeover. Northern natives such as cats, dogs, bears, deer, horses and others made the rainforests and grasslands their home, and to this day they hold this territory. Llamas and tapirs, arguably South America's most iconic animals, are, in fact, descendants of invaders from these times. The most impressive migrants to these lands were the now-extinct sabre-toothed cats. Enlarged canines had first evolved in the Permian gorgonopsids, and this trait had popped up a number of times throughout prehistory, even in herbivores. In the Eocene and Miocene there were feline-like carnivores, cousins of the cats, that experimented with these kinds of teeth, but the true sabre-toothed cats appeared much later. Among the largest of these was *Smilodon populator*, a robust apex predator weighing over 400 kilograms, whose fangs could reach 28 centimetres long. In order to bite properly, *Smilodon* had evolved the ability to open their mouths to over 110 degrees, a gape 45 degrees wider than modern lions. What is remarkable is quite how fragile these teeth were, unable to withstand bending forces, which meant they could not feed like lions or tigers. These modern cats kill their prey by suffocating them, locking their jaws around the nose and windpipe of the victim. *Smilodon* had especially muscled forearms, and scientists now believe that these cats used their body weight to wrestle prey to the ground. Once immobilised, a precise bite would be inflicted through the arteries of the neck, helping prevent damage to the large teeth. Not a wonderful way to go.

In North America, *Smilodon* hunted bison, camels and other large mammals, but those species that moved south during the interchange had a smörgåsbord of potential prey before them. This included *Doedicurus*, a strange armadillo relative that, unlike its modern descendants, reached over 3.5 metres long and the better part of 1.5 tonnes. *Doedicurus* and its relatives (the 'glyptodonts') would have been challenging quarry, even for an adult *Smilodon*. As a defence, their bodies were covered in thick bony scutes that formed a large domed carapace over the back. Scutes on top of the head gave *Doedicurus* its own helmet, and by simply collapsing to the ground it could protect its soft underside effectively. These giant grazers could ward off danger by swinging their club-like tails, which were spiked and grew to a metre long. Nevertheless, fossil evidence shows that at least one juvenile *Doedicurus* did fall victim to a sabre-toothed cat, its skull pierced by massive fangs. In time, both of these species would ultimately fall victim to an even greater threat, as humans spread across the globe.

THE ICE AGE

Since the Eocene, temperatures had been dropping significantly, and by around three million years ago Earth was reaching a tipping point. The South Pole had frozen many millions of years before, and now a combination of low carbon dioxide and changes in the Earth's orbit were causing freezing at the North Pole as well. When the Isthmus of Panama formed, it acted as a bridge for animals moving between North and South America. However, it now blocked

the ocean current that flowed through the Gulf of Mexico, linking the Pacific to the Atlantic. This caused water from the Atlantic to stream northward, bringing more moisture to the North Pole, which then froze. Relatively cold summers meant that snow could survive and not melt, and so it accumulated year after year. In time, this consolidated into ice, which reflected more of the Sun's heat back out into space, causing temperatures to drop further. This had a pronounced effect on baleen whales, which had to travel further for food. Larger animals were able to migrate further to feed, and so it was the smaller species that died out at this time, with only giants like the 30-metre-long blue whale and its relatives surviving, forced to traverse great distances across the planet for food.

By 2.6 million years ago, Earth had entered an ice age, with permanent ice at high latitudes. We are still living in this ice age today, so it is difficult to imagine Earth without its snowy white caps. Ice ages, however, are quite unusual. There have only ever been five, and for 90 per cent of Earth's history there has been no permanent ice at the poles. The extent of this ice has changed with Earth's temperature, which fluctuates up and down about every 40,000 to 100,000 years.

BELOW: Pines are extraordinary cold-climate survivors. In these habitats they grow slowly, making the most of the little light and liquid water available.

This is due to natural changes in the Earth's axis and orbit around the Sun, and some estimates suggest there may have been 50 or more of these swings. The colder times are referred to as 'glacial periods', and the warmer periods are called 'interglacials'. During glacial periods, low temperatures mean that ice sheets can grow further towards the equator – on occasion they have even reached as far south as New York. The ice sheets are not entirely bad news for life, as they can reduce sea levels by well over 100 metres and form land bridges. At the peak of the last glacial period, for example, New Guinea was connected to Australia, and North America connected to Eurasia.

THE HUMAN JOURNEY

Our own genus, *Homo*, originated 2.8 million years ago, just before the Ice Age began, with the earliest fossil evidence being a lower jaw from Ledi-Geraru in Ethiopia. For the next million years or so, a series of species began the transition to modern humans, evolving traits like larger brains, flatter faces and walking fully upright. One of the earliest of these was *Homo habilis*, a metre-tall species which was occasionally eaten by massive 7.5-metre-long crocodiles. The name *Homo habilis* means 'handy man', and there is evidence it was using stone tools like the australopithecines before them.

The greatest tool of all was not a physical object, but the ability to harness fire. Heat breaks down the long molecules of raw food into shorter ones, which are easier for the body to absorb, and so cooking essentially performs a digestive process outside of the body. Cooking also softens tough or fibrous material, further reducing the physical effort needed to break it down. Plant toxins and harmful microbes like bacteria and parasites can also be destroyed by cooking, which makes it safer for us to consume.

At first, humans may have scavenged for fire from lightning strikes or wildfires, treating it like a temporary natural resource. The pivotal moment came when they learned how to make it on demand. Some of the earliest evidence we have of fire use is the way that it changed the anatomy of our ancestors. After *Homo habilis* came *Homo erectus*, around 1.9 million years ago, which had smaller teeth adapted for chewing much softer – perhaps cooked – food. This species was larger and had a much greater muscle mass and brain size than those before, all of which requires more energy to maintain. Our brains today use around a fifth of our body's total energy to operate, and muscle tissue consumes around three times as much energy as fat. *Homo erectus* was also the first hominin to leave Africa, supporting the idea that this species was unlocking more calories by cooking, and using fire as a survival tool as it dispersed. Solid evidence can be found from around 1.5 million years ago, at a site in Kenya called Koobi Fora. Here, scientists have found burned rocks and animal bone in reddened patches of earth, thought to be fire pits. These patches are found alongside stone tools which were probably made by *Homo erectus*.

THE LAST GLACIAL PERIOD

The last cold swing and glacial period began around 115,000 years ago, and during this time a distinct habitat developed in the north. The tundra steppe was a grassy expanse, with small trees, hindered by extreme cold and short growing seasons. By around 15,000 years ago, it was the world's largest habitat, stretching almost around the globe from France to Alaska across the Bering Strait (Beringia). It was inhabited by horses, reindeer, bison and that most iconic animal of the Ice Age, the woolly mammoth. Genetic studies have revealed that the nearest relatives of the mammoth are Asian elephants, but the earliest mammoth ancestors are found in 5-million-year-old South African deposits. These left Africa around 3.5 million years ago, spreading to Romania and China, before one line eventually evolved into the woolly mammoth around 40,000 years ago. Many frozen mummies of mammoths have been

RIGHT: As winter approaches woolly mammoths migrate through the snow to find their primary food of sedges and grass.

found in permafrost, providing remarkable detail about how these animals were adapted to the cold. They had thick layers of hair and fat, and small ears and tails to retain body heat, as well as a special pocket to protect the trunk against the cold. Fossil trackways suggest a female-led social structure, which appears to have been normal for all elephant relatives as far back as the Miocene.

Their predators may have included wolves and large cats, such as cave lions, and they both hunted on the tundra in packs and prides. We know this is the case for cave lions because there are direct human observations of the species walking together, recorded as cave paintings in the Grotte Chauvet, in France. Cave lions were also adapted to the cold, with smaller ears and much thicker fur, which was almost certainly white for camouflage.

THE LONG WALK OF HUMANITY

The earliest fossils of our species, *Homo sapiens,* or 'wise man', are found between 350,000 and 300,000 years ago in Morocco. By around 190,000 years ago, humans may have started to leave Africa, perhaps venturing northwards via the Mediterranean. By around 127,000 years ago, human remains were found in caves of southern China. Until this point, humans may have just been travelling across land, but when they reached the south-eastern shores of Asia, they must have started using boats or rafts. The details are lost in time, but we do know that humans reached Sumatra by around 73,000 years ago. From here, humans carried on across Oceania, before reaching Australia possibly as early as 65,000 years ago. In the far central north of Australia, a sandstone shelter called Madjedbebe preserved stone tools, piles of shells, animal bones, seeds and charcoal – there are even fragments of the pigment ochre, which have wear marks on their surface, suggesting the earliest humans there were painting motifs on the walls.

Those humans that headed north out of Africa via the eastern Mediterranean reached Europe around 43,000 years ago. Particularly hardy humans even headed further north, trekking across the Siberian tundra to the east, 16,000 years ago. Another group headed west across the Bering land bridge to North America, which was crossable at the time. From there, some travelled southwards, following the Pacific coast until they reached Chile, around 14,500 years ago. Others had settled all over North America by around 13,000 years ago and developed more advanced stone tools, which have been found with the remains of the ancestors of Native Americans.

By around 6,000 years ago, or maybe earlier, Plains Indian tribes had developed an ingenious method of hunting. For many millennia, bison or 'buffalo' roamed the Great Plains of North

LEFT: Male cave lions had smaller manes than their modern relatives, and their fur was denser and lighter in colour.

America in vast numbers, possibly 75 million strong. Wolves, hunting in packs, had learned how to drive bison off high ledges to their death. This behaviour was quickly adopted, and perfected, by the human hunters. It was a patient and highly skilled process, involving days of preparation and the slow corralling of the bison for miles before their final scare and stampede over the edge.

INTERGLACIAL WATERWORLDS

BELOW: A river winds through the Greenland ice sheet. As ice melts, its ability to reflect light and heat from the Sun diminishes, a feedback cycle that can cause global temperatures to rise.

Interglacials, those warmer periods of the ice ages, are characterised by a reduction in glaciers, a rise in sea levels, increased rainfall and the expansion of forested areas. As water from the melting ice refills the Earth's hydrological systems, wetland habitats can spread, supporting a huge amount of biodiversity. The expanses of flooded ground are particularly good for freshwater fishes, amphibians and invertebrates, as well as the birds that feed on them. Nesting birds are particularly well supported by these watery habitats, as few predators can reach their nests.

The slow creep of water over land to create freshwater wetlands can provide a bounty of food and habitat for life. However, the flow of water during interglacials is not always this calm or productive. Towards the end of the last glacial period, between 16,000 and 13,000 years ago, a large area of western Montana, in North America, was flooding with meltwater from glaciers to the north. This giant lake was called Missoula, and it covered over 10,000 square kilometres. It was dammed and prevented from draining by an ice wall 900 metres high. After filling for around 60 years, the pressure had built up so much that catastrophic failure was inevitable. When the ice wall broke with explosive force, water escaped at 100 km/h, carrying 200-tonne boulders like they were marbles, and scouring the land for 800 kilometres towards the coast. Within 10 hours it was over, leaving nothing but scars of the torrent, which had swept through and remodelled the landscape. This was not an isolated event, and perhaps forty of these megafloods took place over that 3,000-year interval. Another lake further east, called Agassiz, was fed by glacial meltwater from the same source, and was dammed by ice in a similar way. However, Agassiz was around eight times as large. When its ice wall broke 13,000 years ago,

BELOW: The great ice wall, holding back Lake Missoula, towers over a herd of mammoths in North America.

so much freshwater was released into the Atlantic that it temporarily halted an ocean current carrying warm water northwards. This effect was so abrupt that it even changed the climate, causing cooling and a mini glacial period that lasted 1,000 years.

After this short cold snap had ended 11,700 years ago, Earth entered an interglacial phase, which continues to this day. Earth had undergone significant environmental changes throughout its long history, but it was entering a time of quite extraordinary stability. The last 10,000 years was a period so stable that global average temperatures have only varied by a degree or so. Our hunter-gatherer ancestors lived in small groups, and populations grew and shrank depending on climate and food. The stability of the climate was an opportunity for humans to perfect survival, put down roots and begin building civilizations.

BELOW: First Peoples Buffalo Jump State Park in Montana, USA. American bison have been driven off the cliff edges here for thousands of years.

THE HUMAN PLANET

A combination of climate stability and warm temperatures meant that humans were learning to rely on wild grains. Once patterns in temperature and rainfall became consistent, humans could plan ahead and anticipate productivity. Hunter-gatherers began the process of domestication by planting the seeds of their favourite plants in the best positions for them to grow. This kind of agriculture began around 10,500 years ago, and domesticated grains soon evolved to rely on humans for their survival, just as we did on them. Animals like dogs had been tamed years before to help with hunts, but the climate stability now meant that prey animals could be captured, bred and nurtured just like the plants. Domestication resulted in dramatic increases in population sizes, and soon domesticated species were being taken far away from their natural ranges. What is remarkable is that the species that those early farming communities selected are still sustaining eight billion people today.

Over time, the methods of farming became more advanced and technology improved. The ability for an individual to produce more food than they needed led to a division of labour. Societies of many people with jobs unrelated to food production began to develop, along with social hierarchies, armies and politics. Intelligence bore invention, and the human race rose above the laws and order of the natural world. The technosphere began replacing the biosphere.

OUR EXTINCTION

Between 50,000 and 4,000 years ago, large-mammal species, usually over 40 kilogrammes, were being targeted by human hunters. This peaked during the last glacial period and, to date, around 1,000 species that we know about have been wiped off the face of the Earth since 50,000 years ago. The current extinction event, orchestrated by humans, has seen the disappearance of species at least 100 times the natural background rate. Many experts believe we are living through the sixth of Earth's mass extinctions, and hunting is just one of many causes.

Like the end-Devonian mass extinction, huge dead zones are forming in the oceans as nutrients are washed off the land by deforestation and intensive farming practices.

Like the end-Permian and end-Triassic mass extinctions, carbon dioxide and methane are flooding into the skies, causing global warming and climate chaos all over the world. Wildfires are becoming more common, sea levels are rising, and habitats are being lost by intense changes to their seasons and weather systems. Carbon dioxide is dissolving in the ocean waters, causing the death of shelled creatures and corals. Like the asteroid strike, we are destroying vast tracts of habitat and the life they support with incredible speed.

One possible future for our species is sustainable and sympathetic to the complex biological systems that have nurtured life for billions of years. The other future sees denial and the extinction of our kind. If the time we are living in now is recorded in the fossil record, it will tell a remarkable story. Bottle caps and cigarette filters may survive the fossilisation process, as will the deformed chickens we have bred to eat. Concrete will degrade relatively quickly, metals will rust and plastics will slowly dissolve away. A chemical signature will be preserved in the rock layers of nuclear bombs and toxic metals.

Whatever lifeforms survive us will fill the vacant ecological niches, just as they have done after every mass extinction. As Earth history shows us, life does always muddle through. But what of the far distant future of Earth beyond us?

THE SWANSONG OF THE UNIVERSE

Today, there is around 1.3 billion cubic kilometres of water in our oceans, but it will not be there forever. Water is constantly dragged into our planet's interior by tectonic plates, where it mixes and reacts with magma. About a cubic kilometre of water is lost from our oceans each year in this way. As water vapour travels high up into the atmosphere, it cools and condenses to form clouds and rain or freezes, if it reaches even higher. This is a natural cold trap that prevents the escape of water into space, but our planet is due to get much warmer in the distant future. The Sun, like any main sequence star, will slowly become hotter and brighter, by around 10 per cent every billion years. Eventually, it will be too warm and steamy for the cold trap to work, and water vapour will travel higher without freezing. After radiation breaks it down into hydrogen and oxygen, solar winds – a million tons of protons and electrons thrown out by the Sun every second – will sweep it out into the cosmos.

While the surface of Earth gets hotter, its iron outer core is cooling down and solidifying at over 1,000 tonnes per second. Earth's liquid outer core is responsible for producing its magnetic field, which protects the atmosphere from solar winds. When the liquid outer core freezes and the magnetic field disappears, so too will the atmosphere. The freezing of the outer core also means that the source of heat for plate tectonics will be cut off, and this process will slowly grind to a halt.

As the Earth's surface becomes warmer and water evaporates, the oceans will shrink and become much saltier. Conditions will eventually get so bad that the only habitats left will be

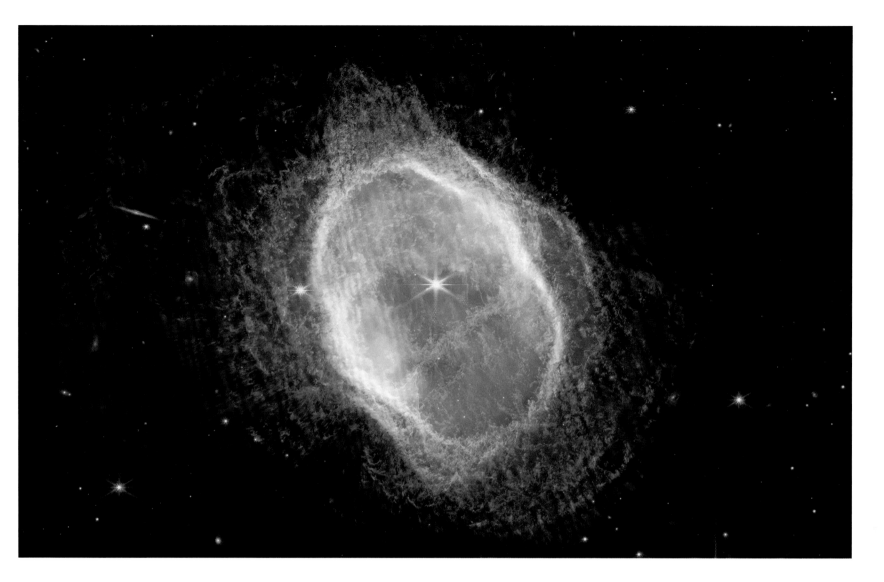

deep caves, the last refuges for the very last lifeforms on Earth – microbes. The end of life on Earth will look remarkably like its beginning, with only bacteria remaining, capable of living amongst extreme heat and toxic chemistry. Even these organisms will vanish in time.

Five billion years from now, the Sun will begin to die. There will be no dramatic end for this star; it is ten times too small to cause a supernova and twenty times too small to produce a black hole. Instead, having exhausted most of the hydrogen in its core, it will begin burning helium. As its core compresses, the atmosphere around it will expand, enveloping and vaporising Mercury, then Venus and, finally, whatever is left of Earth. In five million years, the Sun will have become a red giant and will spend the next two billion years burning what is left of its core in a series of increasingly fruitless chemical reactions. Eventually, the Sun's core will shrink into a white dwarf, heated at first by compression and dimly illuminating the outpouring of gas around it. The final portrait of our solar system will be the wisps of a beautiful new nebula.

ABOVE: The Southern Ring Nebula NGC 3132 as observed by the James Webb Space Telescope in 2022.

BEHIND
THE LENS

Making the documentary series *Life on Our Planet* was a monumental undertaking, involving hundreds of people, 285 shoots in 45 countries, and nearly 2,000 collective days in the field. All the while, thousands of hours of painstaking research and animation were taking place, with experts from all over the world helping to shape the series. From skeletons to eye colour, every detail was pored over, ensuring the very latest science made it to the screen. Experienced natural-history filmmakers worked alongside visual effects (VFX) masters, collaborating to bring the extinct animals and their bygone worlds back to life as truthfully as possible. The first step, back in 2017, even before pen met paper, was to learn about these creatures and their place in the story of our ancient planet. However, this was certainly not going to be easy, and it would take more than 750 people, from myriad organisations, six years to make it happen.

FROM THE GROUND UP

The fossil record is very bad, in the sense that most of the countless life forms that have lived on Earth are not preserved in its rocks. The vast majority of soft and fragile creatures will leave no trace as they decay and, therefore, no proof that they ever existed. There are vanishingly rare instances when the chemistry and conditions are just

PREVIOUS PAGE: Some of the show's many filming locations: Iceland (above), LaPalma (below left), Texas (centre), Chile (centre right), and Malta (below right).

RIGHT: The Danube Delta in Romania is a haven for huge numbers of whiskered terns. This species is well adapted to nesting on water; a unique filming challenge for the crew.

right so that more than just bones and teeth are preserved. Besides these exceptional cases, the chances of anything, even something relatively hard and tough, being preserved are very low. Even bone can be weathered away and recycled back into the ecosystem by scavengers, fungi or bacteria quite rapidly. So the likelihood of a scientist in the twenty-first century happening upon a fossil at all is astronomically low. Thankfully, humans are very curious, and very fond of collecting fossils. Because of this, the amount of material collected and studied to date is still quite impressive. The store rooms of museums and universities are stuffed with all manner of specimens, lovingly curated and conserved for science.

Surprisingly, the most useful and informative fossils are often not much to look at. Take, for example, plankton, which is carefully sieved and washed from ancient muds. Despite being the size of sand grains, and often resembling them too, an expert can discern all sorts of information from these; the shells and fragments can tell you the age of the rock, the productivity of their habitats, the chemistry of the oceans, and even global temperatures. Likewise, the microscopic pollen, which might not steal the focus at a dinosaur dig, can tell you what plants surrounded the giant animals. The rocks around the fossils are equally important, and can tell you a lot about the environment where the creature was buried. Fine muds with a barcode of delicate layers mean it was deposited gently and slowly, in a deep sea or lake, whereas rounded, red sand grains tell you it was an arid and windswept desert. All of this information feeds into reconstructions of past environments, and the stories of the creatures that inhabited them.

Of course, most of what we know about extinct animals comes directly from their fossilised remains. The skeletons can tell you what the animal is related to, how it moved, how it grew and a lot of other information about its behaviour, but the devil is often in the details. Experts will spend decades studying these remains in order to tease out more information, and it may take a number of scientists, and several rounds of study, to do this. The dedication and fascination with these long-extinct species may seem odd, but for perspective it is worth remembering that these fragmentary remains are often all that exists, and may represent a crucial point in the evolutionary map. The *Life on Our Planet* team wanted to tell this story with a number of key species along the timeline, organisms that represented an important innovation, a changing of the guard, or as a demonstration of the extreme adaptations that have evolved over time. Some of these were quite well known, like the dinosaurs, but many more were strange and unfamiliar, including *Lystrosaurus*, an odd, pig-like reptile that survived the end-Permian mass extinction. Perhaps the oddest of these, at least from our modern perspective, was

Anomalocaris, which is unlike anything living today. Bringing this creature back to life required a stack of accumulated research spanning over a century, as a result of working closely with expert scientists.

RESURRECTION

The first stage in recreating the extinct creatures in *Life on Our Planet* was to make the skeletons, using scans and photographs of real fossils, as well as diagrams and discussions with scientists to fill in the gaps. This first skeletal framework provided the points of articulation of limbs, tails and necks, and a solid frame to which soft tissues, such as muscles, can be laid over and attached. For some creatures, such as *Anomalocaris*, *Arthropleura* and the trilobites, the skeleton is external, so the next and final step for those was detailing colour and texture. However, most of the extinct animals in the series were vertebrates, which have internal skeletons and thick layers of muscle, ligament and some kind of covering like hair, feathers or scales. For these creatures the next step was sculpting a body mass and creating the general shape of the animal. This was a crucial

ABOVE: Rows of carefully curated and studied fossils line the stores of the Smithsonian National Museum of Natural History in Washington, D.C.

stage for the scientists to help the artists understand where the flesh would lie, and where to position any extra features, such as crests. In the absence of direct anatomical information, palaeoartists tend to add minimal soft tissue to the skeletons. This 'shrink-wrapping' can make the animals look gargoyle-like and monstrous, so it was important that ours looked healthy. This was the moment when the rough shape of the animal would begin to emerge, just as a sculptor makes a clay model over a wire frame.

As more detail was applied, animators worked to add movement to the models. This began with a simple walk cycle – essentially a catwalk stroll for our animals. Biomechanics experts were consulted throughout as gait and stride length were fine-tuned, as well as the more subtle and characterful movements. Palaeontologists do not have the luxury of observing the movements of extinct creatures, but there is a lot of indirect evidence that we can use. Fossilised trackways, for example, can tell you how the feet landed, the speed of the animal, its size and, in some cases, even interactions with other animals. One of our most powerful tools was observing modern animals and utilising the knowledge of the production team who had collectively spent decades watching wild animals. This was especially useful for those species with living descendants, such as sabretooth cats, mammoths and sharks. However, the most rewarding experiences came when the scientists, animators and filmmakers joined forces to tackle the more unusual creatures, such as *Lystrosaurus*, whose strut was somewhere between that of a mammal and a reptile, and *Anomalocaris*, the strange, mantis, shrimp-like arthropod that glided through the water with a cascade wave of its paddles.

LANDS BEFORE TIME

Once basic movements had been defined with the walk cycle, it was time to release our creatures into their onscreen habitats. One of the biggest challenges was finding locations and avoiding featuring living organisms that had not evolved at the time the scene was set. Grass, for example, is almost impossible to avoid, as are birds in the sky and modern fish in the sea. Since the Cretaceous period, flowering plants have dominated the floras of Earth, making it particularly difficult to find good filming spots for earlier time periods. Likewise in the ocean, finding areas without colourful coral reefs was a real challenge, and often the fish had to be lured away and distracted from the cameras with food. In some cases, Industrial Light & Magic came to the rescue, virtually adding in more appropriate vegetation and painting out troublesome modern species. The Earth has also changed enormously in geographical terms since the eras in which many of these animals were alive. The mountains and oceans are in different places, and even the orbit of the moon has altered, so it was occasionally necessary to tweak the environment to make it more accurate for the period we were portraying. Such is the power of visual effects that the strong features of a location could be used independently, plucked out of the footage and mixed with other elements to make the perfect habitat for the creature.

The wealth of natural history filmmaking experience that the production staff brought to *Life on Our Planet* could not prepare them for filming non-existent creatures. It is quite a sight to see a grown adult waving their arms in the direction of a group of trees, claiming that a dinosaur is walking through them, or a researcher running with a stick, like an amateur pole-vaulter, pretending to be a terror bird. These were the bizarre scenarios that were enacted during the visual effects shoots, which involved getting a whole team to imagine the movements of an animal that would not populate the space until Industrial Light & Magic started working their … well, magic! Even the seemingly simple shots involved all manner of blue screens and practical effects such as rain, smoke and the occasional leaf blower. The biggest challenge was getting a sense of scale, and ensuring the animals could fit into the habitats comfortably. The first time the size of an adult *Diplodocus*, at around 30 metres long, was marked out on location, the crew were noticeably awe-struck by the reality of quite how massive these giants were.

Using the walk cycles, references of modern animals and a healthy dose of imagination, the team went to work filming thin air. It was vital that the VFX animals were treated in the same way as they would be in any other natural history production. For the better part of a century, dinosaurs and other extinct animals were depicted as primitive and unfeeling monsters, but gradually scientists have chipped away at this antiquated perspective, revealing complex life histories and behaviours.

Once the empty habitats had been filmed, some rough animation of the models was drafted on top of the footage to give a sense of the sequence that would follow. This was an opportunity to correct actions and speeds, and also to plan the extra elements such as rocks and plants that needed to be placed around or interact with the creature. This was a very exciting part of the process, and as the blocky models began to get increasingly realistic, it generated a lot of discussion about behaviour. Even with visual references to emulate, choreographing complex behaviours in an organic way is a very involved and skilled process. It is, therefore, particularly impressive that the artists managed to animate complex movements so naturally, with only the guidance of scientists and a few carefully chosen modern animals for reference. The scene where our gorgonopsian stalks and attacks a *Scutosaurus* was both a technical and scientific challenge for this reason. The enlarged canine teeth of the predator were so large they required the mouth to open 90 degrees in order to bite properly. The heavy body armour of *Scutosaurus* left only one possibility – that the gorgonopsian was toppling the prey before inflicting a lethal bite to the neck. The science had pointed us in the right direction, but getting this right, and making it look realistic, was a truly collaborative process.

FEATHERS AND FUR

As the animation progressed, the models had even finer texture added, testing the limits of what fossils can tell us about the detailed appearance of the creatures. Bones, shells and teeth are often the best-preserved parts of the animal, and are therefore a relatively uncontroversial aspect of animal reconstructions. The real challenge for the series was reconstructing the soft tissues – things that rot quickly after the animal has died and do not preserve well, such as muscles, skin, eyes, organs, feathers and fur.

Feathered dinosaurs are not a new concept; in fact, the very first *Archaeopteryx* fossil discovered was a single feather. Even as far back as the 1860s, scientists like Thomas Huxley were making the link between *Archaeopteryx* and dinosaurs. However, this idea did not become popular until much later, when specimens from China began to appear with exquisitely preserved feathers. Unlike *Archaeopteryx,* some of these species were unmistakably dinosaurs, and by the mid-1990s word began to spread that our idea of what dinosaurs looked like had to change. The film *Jurassic Park* had a huge impact on the public's view of dinosaurs as animals rather than monsters, and fuelled interest in what they were really like. The revelation that followed – that many dinosaurs had feathers – was met with a mixture of bemusement and intrigue by the public. In time, the nostalgic visions of reptilian monsters portrayed in the media gradually gave way to more accurate, but decidedly less awe-inspiring, creatures. The production team on *Life on Our Planet* wanted to reignite that awe and fascination by combining scientific accuracy with all the tools of natural history filmmaking. It was critical to make the animals behave as realistically as possible, and to explore what feathered dinosaurs were capable of.

Our oldest feathered species featured in the series is *Anchiornis*, a crow-sized ancestor of birds that demonstrates a transitional stage in evolution between simple gliding and true powered flight. The preservation of *Anchiornis* fossils are so good that scientists have been able to map out the different feather types across its body. The wings and legs had stiff and modern-looking flight feathers, while the body was covered in more of a shaggy fur, and there was even

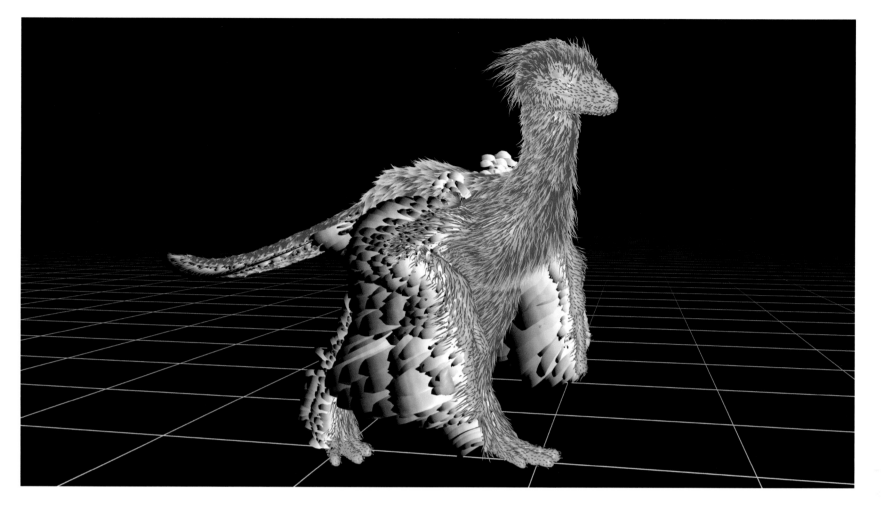

ABOVE: The groom stage of model construction is an intricate and time-consuming process. Here the feathers of *Anchiornis* are taking shape after detailed consultation with scientists.

a crest of feathers on its head. The detail is even more remarkable when these fossils are viewed under the microscope, when tiny structures called melanosomes can be seen. Melanosomes are around one-thousandth of a millimetre across and contain the pigment melanin, which occurs in two forms: eumelanin produces black and brown colours, while pheomelanin gives yellows and orange colours. While those colours may be long gone from the fossils, the shape of the melanosomes remains, and this reveals what kind of melanin was contained within them. We know that eumelanin is found in oval-shaped melanosomes, whereas pheomelanin is found in more spherical melanosomes, so by looking at the concentrations of these different melanosomes, scientists were able to reconstruct the patches of black and white colour over *Anchiornis*'s body, including its rufous head crest.

Feathers of some description have now been discovered in a good number of Chinese dinosaurs, but most species do not preserve their detail. So, with just bones, how can you tell if a dinosaur had feathers? One line of evidence comes from skin impressions, which are sometimes found in the rock surrounding the bone. Skin-impression fossils helped the artists at Industrial Light & Magic create textures for *Triceratops* and *Diplodocus*, and there is even an entire fossilised front foot of *Edmontosaurus*. In the case of *Tyrannosaurus*, the only skin impressions show a scaled skin, but this is not the end of the story. A powerful tool for reconstructing any extinct animal is called 'bracketing', which involves using the ancestors and descendants of a species to piece together what they looked or behaved like. In the case of *Tyrannosaurus*, we have fossils of ancestors from around 60 million years earlier with feathers, and, of course, birds alive today also have feathers. The last thing to consider is that many scientists believe the large size of *Tyrannosaurus* meant that it did not need insulation, and that it would, in fact, overheat with a feather coat. As with all things, the truth probably lies somewhere in between, and in the series the best evidence supported a few quill-like filaments. Modern birds often have strong differences between the males and females, so feathers were added to the top of the male's neck as a sexual display.

However, the way scientific discussion works is with publication of information, followed by counter arguments backed with even more data. With more-popular species, this scientific discourse can get quite animated, and no other animal demonstrates this quite as well as *Tyrannosaurus rex*. The disproportionate research attention given to *T. rex* is understandable, but it does mean that every detail about the animation in this series will, undoubtedly, be scrutinised with merciless enthusiasm. Of course, it is impossible to satisfy everyone, but the science always led the way. In addition to group hunting and the feather coverage, the decision to cover the teeth of *T. rex* remains a controversial topic. Lips, or some other type of soft tissue covering the teeth, may be important for preventing desiccation and cracking, or perhaps to protect them in some other way. Whatever their function, lips are much more common than bare teeth in modern animals, even those with very large teeth like hippos, Komodo dragons, snakes, monkeys and many others. These kinds of decisions were difficult because the more accurate reconstructions are often less scary or impressive than the traditional ones. However, it was important to depict iconic species like *T. rex* as animals and not monsters, and so every opportunity was taken to show more complex and intimate social behaviour. Evidence for *Tyrannosaurus* courtship comes from modern birds and crocodiles, where the male often displays to the female, demonstrating their physical fitness with calls and dances. Fossilised scrape marks made by theropods like *T. rex* have been found in Colorado, USA, which show that animals like this displayed in much the same way as modern birds. The exact dance routine is lost in time, but it was fun to try to reconstruct this behaviour using modern bird references, and scaling them up to work for 12-metre-long dinosaurs.

Scientific discoveries came thick and fast during the production of *Life on Our Planet*, and the team were in constant contact with experts to make sure our models were up to date. In

RIGHT (ABOVE): Empty 'backplate' footage, filmed in Morocco.

RIGHT (BELOW): The same image populated with a herd of *Theosodon*, courtesy of the talented artists at Industrial Light & Magic.

2018, debate had been ignited about branching structures found in Chinese pterosaurs, which the authors suggested were feathers. This was quite a dramatic change from the scaly-skinned pterosaurs that had been depicted previously, but the experts were convinced, and science had to lead the way. In 2022, the debate was settled by an international team, who had not only found feathers on the crest of a pterosaur from Brazil, but also a variety of melanosomes as well.

Hair was equally tricky to pin down in some of our extinct mammal species. Thanks to the bracketing technique we could be fairly sure that sabretooth cats and *Megacerops* had fur, and indeed mummified hair is preserved directly for mammoths. But what about the distant Permian ancestors of mammals? There is nothing like the gorgonopsian alive today, and no exceptional fossils that preserve the skin, so in this case, evidence came from lumps of fossilised faeces, or coprolites, found in Russia from rocks the same age as our gorgonopsians. By analysing the shape, size and contents of the droppings, scientists have concluded that they were produced by a large carnivore with a fast metabolism. Within the lumps were not only undigested teeth and bones, but also hairs, suggesting there were furry animals living at the time. Gorgonopsians were amongst the most advanced large predators of the mammal line around, so this all points towards them having hair — by far the oldest record of this adaptation in the fossil record.

BELOW: The animation process of the *Tyrannosaurus rex* mating dance was highly complex, and required perfectly timed movements to look natural and realistic.

NEXT PAGE: Extra elements like footprints, vegetation, smoke and dust were a combination of practical and visual effects. Here a mammoth herd cast long shadows in the trodden snow.

PAST MEETS PRESENT

Aside from the visual effects, a major component of the series was showcasing modern natural history and the biodiversity of the world today. Every organism on Earth has its own unique journey, stretching back to that very first single-celled lifeform, LUCA. Celebrating the living descendants alongside their progenitors was a unique opportunity to tell those stories, and to break down the divide between the past and the present.

Normally, the natural history filming would offer a pleasant respite for the team, who have worked in the industry for years, but nobody could have predicted what was to come. The Covid-19 pandemic hit at just the right time to disrupt or entirely write-off shoots that had been planned for months, if not years. While nations around the world locked down, the team scrambled to find new locations or alternative routes into countries. The unsung heroes of these kinds of programmes are the production coordinators, who, by some small miracle, managed to get everyone to where they needed to be in a time of relentless chaos. Indeed, everyone had to adapt to radically different working conditions, including heaps of paperwork and endless tests, long and solitary quarantines in hotels, rapidly changing guidelines and a team exhausted by the stress. This was no ordinary series, in any respect, but *Life on Our Planet* found a way, and that will always remain a source of immense collective pride.

Once the crew were on the ground and in the field, they could focus on what mattered: the incredible wildlife fostered by our planet. All of the natural history sequences were linked to crucial stages in evolution, and so the goal was to film the species that epitomised those concepts. For example, what better animal to film for the evolution of jaws than the sarcastic fringehead, a bizarre fish living off the coast of California, which has a huge mouth and aggression to match, despite only being 30 centimetres long. The divers had to brave cold water and strong currents, scouring the sea floor for days just to find them. The next step was getting the bulky camera equipment into position, including a long probe with a prism at the end, by lowering it carefully into the fringehead's rocky arena. One wobble and the picture would be out of focus; the divers had to hold on tight, fighting both the currents and an over-familiar bass attracted to a camera operator who apparently resembled a female fish. On the other side of the Pacific, just north of Sydney, Australia, the crew filmed the beautiful peacock spiders to demonstrate how advanced modern arthropods can be. Even with special lenses, and scientists on hand to advise, waiting for the males and females to do their thing was a unique challenge. At about the size of a grain of rice, these little gems were tricky to keep in focus, especially given their fondness

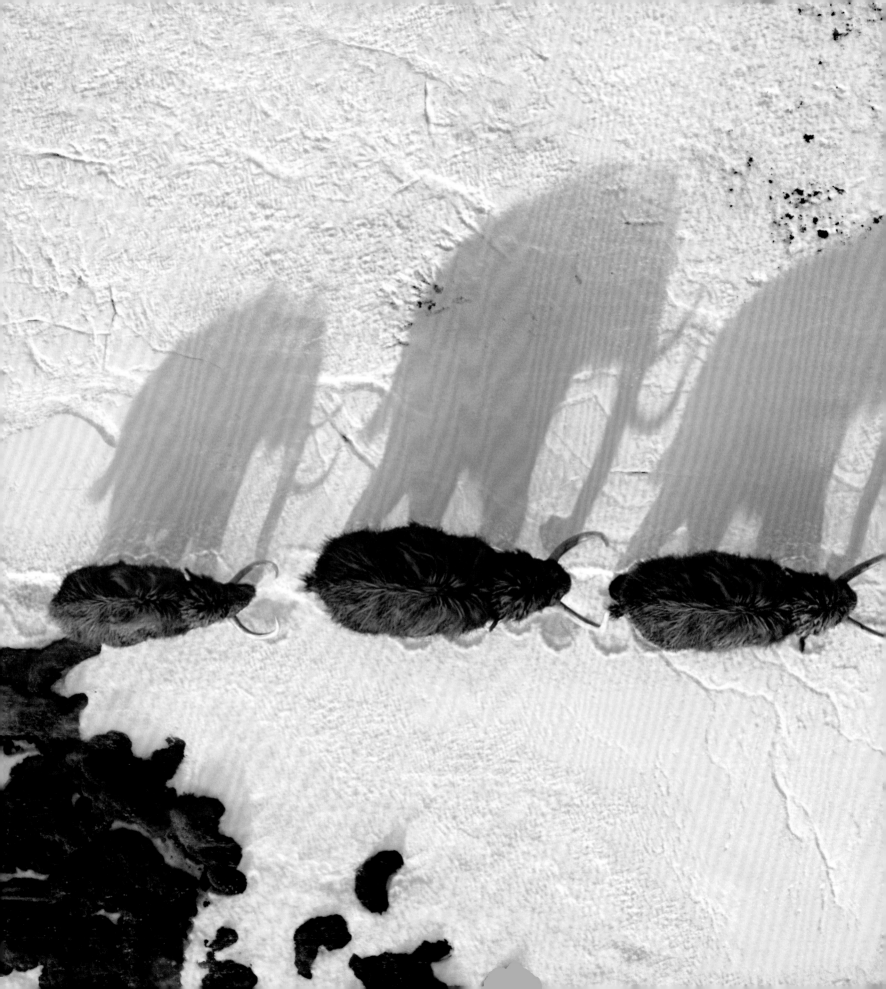

for jumping away. Patience is the critical ingredient for natural history shoots, and getting the perfect sequence takes a unique dedication ... and focus.

Just as much research went into the natural history shoots as the visual effects, which were also an opportunity for scientists to be more involved during filming in the field. Their expertise and experience was crucial in helping to find the animals and provide information about their behaviour that we could capture on film in interesting ways. The scientists, too, benefited from seeing the footage and the incredible detail revealed by filming techniques such as slow motion and macro photography.

SCORE AND FOLEY

In the final stages of production, a truly unique programme began to emerge as natural history footage was blended with the visual effects sequences. Such a grand and dramatic story needed a score to match, and composer Lorne Balfe joined the team to meet the challenge.

The music had to feel timeless and match the grandeur of the four-billion-year story we were telling, which involved some quite unconventional instruments. Among these was a 3D printed replica of the Divje Babe flute, discovered in Slovenia in 1995. It was originally carved out of cave bear bone by a Neanderthal, and remains the oldest instrument ever discovered, dating to around 55,000 years ago. Ultimately, hundreds of layers of audio were woven together to create the ancient soundscapes, including orchestral pieces that brought everything together. When the time came for the orchestra to record their parts, it was striking to consider how many artists from completely different worlds had worked together on this project; all speaking the unique languages of their particular industry. Scientists have a lexicon of obscure terminology and Latin names, camera operators speak in millimetres, stops and framerates, while musicians communicate in a wonderful foreign language of bars, phrases and quaver rests.

One of the final layers was the foley – a process of adding the various grunts, rasps, thuds and splashes that the animals make as they move around on screen. With a huge digital library and practical techniques at their disposal, the foley artists could also add sound texture to make the world feel more natural. If successful, the foley is something that no one really notices, instead adding an intangible sense of realism to a scene.

Adding modern wildlife sounds is straightforward because they can be recorded in the field, but crafting the vocalisations of long-extinct animals is a lot trickier. The best evidence we have for the kinds of sounds that extinct animals made comes from studying the hollow spaces in the skulls where the inner ears once sat. For example, *Allosaurus* has an inner ear very much like a crocodile, and so were probably best adapted to hearing low frequencies. On the other hand, dinosaurs with a longer region of the inner ear, called the cochlea, could better hear high-frequency sounds, and these are found in some of the earliest dinosaur relatives. Scientists believe this may have evolved to hear the high-pitched chirps of their hatchlings. Aside from physical evidence, the bracketing method can also be helpful. The closest living relatives of dinosaurs are birds and crocodiles, so there is a spectrum of vocalisations to choose from. For example, the cassowary today is known to make low-pitched bellows that travel effectively through the thick vegetation of its Queensland rainforest habitat. Crocodiles, on the other hand, hiss with aggression, and occasionally rumble to communicate through water. The simple answer is that we will never really know what these extinct animals sounded like, and there is creative fun to be had in exploring the possibilities.

LIFE ON OUR PLANET

We are living through a revolution in palaeontology, with frenzied interest in answering some of the biggest questions. In the last few years, exciting discoveries and new techniques have revealed details about the history of life that we did not believe were possible to answer. Long-standing debates have been settled, scores of new species have been described, and powerful new technological advances have been applied in the most incredible ways. One central character stands alone in the series. The greatest of them all, which has at once nurtured and ruled over life for billions of years, is the Earth itself. The story of life is driven forwards by the shifting of continents, the volcanic fireworks, the rising and falling of seas, soaring and plummeting temperatures, and the complex dance of chemistry in the oceans, on land and in the air.

As our knowledge of the Earth's past grows at an accelerated pace, so too does our reflection of its future. The fossils and rocks are a four-billion-year record of the huge changes that have taken place. They are solid evidence of mass extinctions and climate chaos, with the all-too-familiar signatures we recognise today. If we ignore the warnings from prehistory – the lessons written in stone – the future is bleak for life on our planet, but hope springs from those same rocks. We are adaptable and intelligent custodians of this world, with the tools and resources required to heal it and reverse the destruction. It would, and should, be our greatest triumph.

NEXT PAGE:
The North American sabre-toothed cat *Smilodon gracilis*, brought back to life after 2 million years.

ACKNOWLEDGEMENTS

The production of *Life on Our Planet*, like the story of Earth itself, was long, complicated, and often very challenging. It was only made possible thanks to an amazing team of people, united by a shared passion to make a series with a scale and ambition never attempted before. The team numbered more than 500 people, and included researchers, animators, scientists, editors, producers, cinematographers, field assistants, coordinators, managers, accountants, musicians, colourists, audiophiles and so many more. Their combined expertise, dedication and willingness to collaborate all helped make the series become as ground-breaking as its subject matter. Moreover, the resilience shown by everyone to always go *above and beyond* despite an unprecedented global pandemic was, quite simply, remarkable.

The production team at Silverback Films would particularly like to express their thanks to the teams at Netflix (most notably Adam, Sara, Cameron, Karen and Oli) and Amblin Television (especially Darryl and Justin) for their guidance, support and trust. We are hugely grateful to the artists at Industrial Light & Magic (lead by Jonathan and Louise). The concentration of talent and passion for prehistoric creatures at ILM was extraordinary, and it was a pleasure to work with them all. Thanks too to Lorne Balfe and his team for creating a such a timeless and enchanting score. We will forever be grateful to our scientific advisors at Yale, Bristol and beyond, who helped bring this series to life as accurately as possible. In addition, the author would like to thank Laura Barwick, Michael Bright, Phoebe Lindsley and Dan Tapster who helped elevate this book to mirror the production value of the series.

CREDITS FOR SILVERBACK FILMS

SERIES PRODUCERS
Dan Tapster
Alastair Fothergill
Keith Scholey

LINE PRODUCER
Fiona Marsh

PRODUCERS
Adam Chapman
Sophie Lanfear

Barny Revill
Nick Shoolingin-Jordan
Gisle Sverdrup

ASSISTANT PRODUCERS
Joe Fereday
David Heath
Katie Parsons
Helen Sampson
Jolyon Sutcliffe
Amy Thompson

Loz Whittaker
Darren Williams

RESEARCHERS
Bertie Allison
Edd Dyer
Frank Jordan
Ida May-Jones

JUNIOR PRODUCTION MANAGER
Stacey Hill

PRODUCTION COORDINATORS
Lisa Connaire
Nicole Hobart
Anna Kington
Emily Jayne Turner

EDITORS
Alex Boyle
Rob Davies
Charlie Dyer
Andy Netley
Dave Pearce

Sam Rogers
Rupert Troskie

SCRIPT SUPERVISOR
Patrick Makin

VFX PRODUCER
Clare Tinsley

PICTURE CREDITS

a = above   b = below
l = left   r = right
c = centre

2–3 Silverback Films/ Industrial Light & Magic; **4–5** Silverback Films/ Elastic; **6–11** Silverback Films/Industrial Light & Magic; **13** Joe Fereday; **14** Silverback Films/Industrial Light & Magic

1 ORIGINS

16–17 Christophe Boisvieux/Getty Images; **18–19** Jamie Pham/ Alamy; **20–21** Michael Schwab/Getty Images; **22** Silverback Films/ Industrial Light & Magic; **24** Jeff Vanuga/naturepl. com; **25–6** Silverback Films/Industrial Light & Magic; **28–9** William D. Bachman/Science Photo Library; **31** Alex Mustard/ naturepl.com; **33** Konrad Wothe/naturepl.com; **34–5** Martin Molcan/ Alamy; **37** Arctic Images/ Alamy; **39** BIOSPHOTO/ Alamy; **40** Marek Mis/ Science Photo Library; **42** Remi Masson/naturepl. com; **44–5** Silverback Films/Alastair MacEwen; **47** Silverback Films/ Eduardo Ferrer; **48** Norbett Wu/Minden/ naturepl.com

2 THE FIRST FRONTIER

50–1 Silverback Films/ Doug Anderson; **52–3** Zeytun Travel Images/ Alamy; **55** Birgitte Wilms/ Minden/naturepl.com; **56–7** Silverback Films/ Hugh Miller; **59** Silverback Films/Doug Anderson; **60–1** Ralph Pace/Minden/ naturepl.com; **62–3** Tui De Roy/naturepl. com; **65** Juergen Freund/ naturepl.com; **67** Sinclair Stammers/Science Photo Library; **68–9** Silverback Films/Doug Anderson; **71–3** Silverback Films/ Industrial Light & Magic; **74** Silverback Films/ Thorben Danke; **75** Alex Hyde/naturepl.com; **77–8** Silverback Films/Industrial Light & Magic; **81** Alex Mustard/naturepl.com; **83** Emanuele Biggi/ naturepl.com

3 INVADERS OF THE LAND

84–7 Silverback Films/ Industrial Light & Magic; **89** Millard H. Sharp/ Science Photo Library; **90–1** Silverback Films/ Industrial Light & Magic; **93** Juergen Freund/ naturepl.com; **95** Stephen Dalton/naturepl.com; **96–7** Remi Masson/

naturepl.com; **98–112** Silverback Films/Industrial Light & Magic

4 IN COLD BLOOD

114-5 Silverback Films/ Jamie McPherson; **117** Fabio Liverani/naturepl. com; **118–121** Silverback Films/Industrial Light & Magic; **123** Huw Cordey/ naturepl.com; **124–5** Silverback Films/Industrial Light & Magic; **127** John Wollwerth/Alamy; **129** John Sirlin/Alamy; **130** Carver Mostardi/Alamy; **131** Design Pics Inc/ Alamy; **133** Tim Laman/ naturepl.com; **134** Ann & Steve Toon/naturepl.com; **136–9** Silverback Films/ Industrial Light & Magic; **140** Silverback Films; **143** Alex Mustard/ naturepl.com; **144** Silverback Films/Industrial Light & Magic

5 THE RISE OF THE DINOSAURS

146–7 Silverback Films/ Industrial Light & Magic; **149** Edwin Giesbers/ naturepl.com; **150** Silverback Films/Industrial Light & Magic; **152** Dirk Wiersma/Science Photo Library; **154–7** Silverback Films/Industrial Light & Magic; **158** Jorge León

Cabello/Getty Images; **159** Natural History Museum London/Science Photo Library; **161** Pawel Opaska/Alamy; **162** Nick Hawkins/naturepl.com; **164–5** Richard Bouget/ AFP via Getty Images; **166–7** Silverback Films/ Industrial Light & Magic; **169** Staffan Widstrand/ Wild Wonders of China/ naturepl.com; **170–7** Silverback Films/Industrial Light & Magic; **178** imageBROKER/Alamy; **181–3** Silverback Films/ Industrial Light & Magic

6 PARADISE LOST

184–5 Silverback Films/ Dane Bjerno/Industrial Light & Magic; **186–7** Scott Leslie/naturepl.com; **188** Piotr Naskrecki/ naturepl.com; **190t** Silverback Films/Pete Cayless; **190b** Silverback Films/Alex Falk; **193** Silverback Films/Industrial Light & Magic; **194** Silverback Films/Pete Cayless; **196–7** Piotr Naskrecki/naturepl.com; **199** imageBROKER/ Alamy; **201** Klein & Hubert/naturepl.com; **203** Silverback Films/ Industrial Light & Magic; **205** Brandon Cole/ naturepl.com; **206–221**

Silverback Films/Industrial Light & Magic

7 THE LONGEST WINTER

222–3 Silverback Films/ Dane Bjerno; **224–5** Silverback Films/Industrial Light & Magic; **226** Floris van Breugal/naturepl.com; **229** Bruno D'Amicis/ naturepl.com; **231** Paul Williams/naturepl.com; **232–3** Silverback Films/ Gavin Thurston; **234** John Cancalosi/naturepl. com; **236** Tom Stack Assoc./Alamy; **239** Konrad Wothe/naturepl. com; **240–1** Mark Carwardine/naturepl. com; **242–3** Silverback Films/Barrie Britton; **244** Peter Scoones/naturepl. com; **246–9** Silverback Films/Industrial Light & Magic; **251** Feng Wei Photography/Getty Images; **252** Gerry Bishop/Alamy; **253** Robert K. Chin/Alamy; **255** Silverback Films/Industrial Light & Magic

8 THE AGE OF FIRE AND ICE

256–7 Silverback Films/ Industrial Light & Magic; **259** David Tipling/ naturepl.com; **260–1** Silverback Films/Industrial

Light & Magic; **263** Lou Coetzer/naturepl.com; **265** Cultura Creative RF/ Alamy; **266–7** Silverback Films/Industrial Light & Magic; **269** Silverback Films/Gavin Thurston; **270–3** Silverback Films/ Industrial Light & Magic; **274** Silverback Films/ Jamie McPherson; **276–8** Silverback Films/Industrial Light & Magic; **280** Uri Golman/naturepl.com; **281** Silverback Films/ Industrial Light & Magic; **282** Silverback Films/ Jamie McPherson; **285** NASA,ESA,CSA,STScl/ Science Photo Library

9 BEHIND THE LENS

286–7a Jamie McPherson; **286bl** Joylon Sutcliffe; **286cr** Duncan Graham; **286–7b** Katy Fraser; **287cl** Jamie McPherson; **288–9** Sophie Lanfear; **291** Chip Clark, Smithsonian Institution; **295** Silverback Films/ Industrial Light & Magic; **297a** Silverback Films/ Jamie McPherson; **297b** Silverback Films/Industrial Light & Magic; **298–305** Silverback Films/Industrial Light & Magic

COVER Silverback Films/ Industrial Light & Magic

Tom Fletcher PhD is a vertebrate palaeontologist and wildlife expert, specialising in sharks and fossil fishes. He completed his masters at the University of Bristol, and his PhD at the University of Leeds. As an academic he has published a variety of scientific papers, and continues to collaborate internationally as an Honorary Research Fellow of Palaeobiology at the University of Leicester. Tom now works at Silverback Films as a scientific adviser for a number of big-budget natural history series.

Silverback Films is a world-leading producer of natural history films for both television and cinema. Founded in 2012 by Alastair Fothergill and Keith Scholey, it brings together an exceptional team of wildlife filmmakers and has produced shows and features that are among the most successful ever created - including *Our Planet* (Netflix), *The Hunt* (BBC), *A Perfect Planet* (BBC), *The Mating Game* (BBC), seven Disneynature features (Disney+), *Wild Isles* (BBC) and most recently *Life on Our Planet* (Netflix).

Copyright information for all illustrations can be found on p.311
Publishing Director: Albert DePetrillo
Project Editor: Phoebe Lindsley
Editor: Michael Bright
Picture Research: Laura Barwick
Image Grading: Stephen Johnson, www.copyrightimage.com
Design: Richard Atkinson
Production: Antony Heller

Published by Sourcebooks
P.O. Box 4410, Naperville, Illinois 60567-4410
(630) 961-3900
sourcebooks.com

Originally published in 2023 in Great Britain by Witness Books, an imprint of Ebury Publishing. Witness Books is part of
the Penguin Random House group of companies whose addresses can be found at global.penguinrandomhouse.com.

This book is published to accompany the television series entitled *Life on our Planet*,
produced by Silverback Films and first broadcast by Netflix in 2023.

Cataloging-in-Publication Data is on file with the Library of Congress.

Printed and bound in China by C&C Offset Printing Co., Ltd
CCO 10 9 8 7 6 5 4 3 2 1